Toxins and Cancer Therapy

Toxins and Cancer Therapy

Editor

Adam E. Snook

MDPI • Basel • Beijing • Wuhan • Barcelona • Belgrade • Manchester • Tokyo • Cluj • Tianjin

Editor
Adam E. Snook
Jefferson University
USA

Editorial Office
MDPI
St. Alban-Anlage 66
4052 Basel, Switzerland

This is a reprint of articles from the Special Issue published online in the open access journal *Toxins* (ISSN 2072-6651) (available at: https://www.mdpi.com/journal/toxins/special_issues/toxins_cancer).

For citation purposes, cite each article independently as indicated on the article page online and as indicated below:

LastName, A.A.; LastName, B.B.; LastName, C.C. Article Title. *Journal Name* **Year**, *Volume Number*, Page Range.

ISBN 978-3-0365-0190-1 (Hbk)
ISBN 978-3-0365-0191-8 (PDF)

Contents

About the Editor

Adam E. Snook, Ph.D., is an Assistant Professor in the Department of Pharmacology and Experimental Therapeutics at Thomas Jefferson University. He received a B.S. in Pharmacology and Toxicology (2001) from the University of the Sciences and a Ph.D. in Immunology and Microbial Pathogenesis (2008) from Thomas Jefferson University. Dr. Snook was a founding member and served as the Director of Antibody Development for five years at Invisible Sentinel Inc., a leading molecular solutions company specializing in food safety and a current subsidiary of bioMérieux. He joined the faculty at Thomas Jefferson University in 2013 in the Department of Pharmacology and Experimental Therapeutics, where he is studying the mechanisms underlying colorectal cancer tumorigenesis and the interaction between cancer and the immune system to develop new options to prevent or treat gastrointestinal cancers. His work has led to seven investigator-initiated clinical trials examining colorectal cancer chemoprevention, cancer vaccines, and CAR T-cell therapies. Dr. Snook has authored over 80 book chapters and papers in prestigious journals, including *Cancer Research, Gastroenterology, Journal of Clinical Investigation, Journal for Immunotherapy of Cancer (JITC), Cancer Immunology Research, and Cancer Cell*, and his work has been featured in Nature Outlook, US News and World Report, Forbes, Reuters, the Washington Post, the New York Times, the Philadelphia Inquirer, and others.

Editorial

Mobilizing Toxins for Cancer Treatment: Historical Perspectives and Current Strategies

Jessica Kopenhaver, Robert D. Carlson and Adam E. Snook *

Department of Pharmacology and Experimental Therapeutics, Thomas Jefferson University, 1020 Locust Street, Philadelphia, PA 19107, USA; Jessica.Kopenhaver@jefferson.edu (J.K.); Robert.Carlson@jefferson.edu (R.D.C.)
* Correspondence: Adam.Snook@jefferson.edu; Tel.: +1-215-503-7445

Received: 15 June 2020; Accepted: 17 June 2020; Published: 23 June 2020

The level of complexity in a disease like cancer presents a number of challenges for effective treatment development, which require significant innovation to overcome. Enthusiasm for immunotherapies and other types of biotherapeutics has grown substantially over the past decade, as additional insight into the interplay between tumors and the immune system has allowed for a departure from harsher conventional systemic treatments. However, amidst these impressive advances, these therapies may still fall short for many patients. Faced with this dilemma, more biotherapeutic options continue to be researched as potential primary or adjuvant treatments.

For millennia, poisonous compounds have been used for medicinal purposes such as mild pain relief or numbing during surgery. Even in the modern age, plant and animal toxin-derivatives continue to be widely used as treatments for a variety of ailments. The anticoagulants tyrofabin and hirudin, for example, originate from venom of the African saw-scaled viper and leech secretions, respectively [1]. Even pathogenic bacteria typically considered harmful to healthy tissue may prove to be clinically useful as studies have shown that toxins produced by these organisms can be manipulated to target aberrant cells in a tissue- or cell-specific manner [2–5].

In this Special Issue, we explore how toxins may be used as powerful treatments against certain cancers. The compiled articles cover how naturally-derived poisons can be utilized for cancer therapy on multiple levels, from interrogating cytotoxic pathways in different cell types, to exploiting toxic derivatives for pain relief in patients suffering from radiation sickness [3,5]. The Special Issue presented this month helps to expound upon this field of research and demonstrate the potential for its clinical applicability.

One of the first uses of toxins as cancer treatments dates back to the early twentieth century, most notably by William Coley, a bone surgeon who discovered that a combination of heat-killed and systemically administered bacteria could shrink osteosarcomas [6,7]. The inception of cancer immunotherapy can arguably be traced back to the innovation of "Coley's Toxins," which initiated queries into how a patient's immune system can be triggered to kill cancer cells [2]. Immunoediting, a prominent idea in the field of immunotherapy [8], asserts that while the immune system is at first able to recognize and kill portions of cancer cell populations, the cancer gradually develops mutations that permit evasion of immune detection, allowing for tumor growth and eventual metastasis [2]. Over the past several decades, clinical strategies to overcome stagnancy in cytotoxic T cell or NK cell responses include utilizing immune checkpoint inhibitors, such as PD-1/PD-L1 or CTLA-4 blockade [9–12], or direct infusion of cytokines like IFNα [13,14]. Moreover, vaccines against neoantigens [15] and known tumor-associated antigens, could prove useful for patients with genetic predispositions to cancer. Currently, there is a phase II clinical trial investigating the ability of a mucin 1 (MUC1) vaccine to prevent adenoma recurrence in patients at high-risk of colorectal cancer [16]. Oncolytic virotherapy, like the FDA-approved talimogene laherparepvec ("T-VEC") [17,18], is yet another instance of a therapeutic derived from bioengineering. Remarkably, this kind of virotherapy works to reshape and adapt the tumor microenvironment (TME) to boost immune infiltration.

The mechanism of action for such biotherapeutics must be well understood for effective employment of the treatment, as one study considers. Shiga toxins (Stxs) produced by *Escherichia coli* and *Shigella dysenteriae* 1 pathogenic bacteria bind to the cell surface receptor glycosphingolipid globotriaosylceramide (Gb3) [19] and induce apoptosis by inhibiting protein synthesis [3]. Gb3 is highly upregulated in Burkitt lymphoma (BL) cells [20], and Stx/verotoxins, VT-1 and VT-2, have been used in several preclinical studies, albeit with little success due to abundant cytotoxicity and poor understanding of verotoxin-induced apoptosis [21]. Detailed in one paper, treating BL cells with VT-1/Stx1 consistently induces the endoplasmic reticulum (ER) stress response by activating ER stress sensors, IRE1 and ATF6, as well as increasing expression of the transcription factor C/REB homologous protein (CHOP) that normally signals for programmed cell death. The role of VT-1 in cell death is noted to be cell-specific, and in fact may shield certain tumor cells from death instead of inducing apoptosis. ER stress enhances VT-1-induced apoptosis through CHOP in BL2 cells, but not in Ramos cells [3]. Strikingly, VT-1-induced ER stress triggers ER-phagy that in turn restrains apoptosis in Ramos cells.

Escherichia coli protein toxin, cytotoxic necrotizing factor 1 (CNF1), acts as an effective anti-neoplastic in glioma mouse models, reducing tumor volume and increasing survival, all while preserving the functional properties of the surrounding neurons [22,23]. As one paper acknowledges, therapies against glioma cells must be able to cross the blood-brain barrier (BBB), otherwise, treatment would have to be directly injected into the brain [4]. To circumvent invasive cranial injections, CNF1 was reengineered with an N-terminal BBB-crossing tag. Not only does this BBB-CNF1 variant, referred to as the An2-CNF1-H8 variant, show comparable activity to its wild-type (WT) counterpart, but it is also able to be purified in native conditions. This variant also exerts cell growth arrest of U87MG GBM cells in a similar fashion to unmodified WT-CNF1 and upregulates pro-apoptotic protein Bax expression. Experiments performed on endothelial cells demonstrate that the An2-CNF1-H8 variant is able to enter cells and perform its intended functions, as indicated by equivalent actin architecture changes to that of the WT. Intravenous administration of the An2-CNF1-H8 variant upregulates spinophilin in the mouse hippocampus, suggesting BBB bypass. Altogether, these results demonstrate that the An2-CNF1-H8 variant is likely able to cross the BBB to induce cell death in GBM cells [4] and may be translated to future clinical studies.

Similar to clinically-approved CD3-based bispecifics, some immunotoxins utilize antibody-like specificity to recognize tumor antigens, while also possessing a toxic domain that releases a toxin into the target cell following internalization [24,25]. One study explores how immunotoxin efficacy could be improved by modulating the intracellular trafficking of the toxin [26]. The inclusion of a furin cleavage site allows immunoconjugates derived from RNase T1 and the fungal ribotoxin α-sarcin (scFvA33furT1 and IMTXA33furαS, respectively) to be purified with optimized properties for colorectal tumor treatment. It is also noted that the two immunotoxins are trafficked in different pathways after endocytosis. After binding to their target GPA33 on the surface of W1222 colorectal cancer cells, IMTXA33furαS goes through the endosome-Golgi-apparatus network, and scFvA33furT1 appears distributed between the lysosomes and the Golgi-apparatus. The differences in trafficking pathways between the two immunoconjugates align with what is observed from their original constructs [27]. In vitro functional characterization of these variants demonstrates enhanced antitumor efficiency due to increased ability to release their toxic domain into the cytosol, as well as high thermostability and target specificity.

Aside from direct applications as cancer treatments, toxins could be used to mitigate pain directly caused by tumor pressure, or neuropathic pain as a side effect of radiation or surgery in cancer patients [5]. Many clinical studies [28–38] have investigated the use of botulinum neurotoxins (BoNT) as potent systemic analgesics, as these toxins block acetylcholine release from the neuromuscular junction or inhibit neurotransmitters at both peripheral and central sensory levels [39–42]. Additionally, some sources claim that spiking certain cancer cell lines with BoNT slows growth and mitosis, as well as enhances apoptosis [43]. Studies of pain induced by radiation and/or surgery suggest that the local injection of BoNT improves neuropathic pain and local muscle spasm in the direct vicinity of the site of surgery and/or radiation. However, this type of pain-management therapy requires blinded and placebo-controlled studies to confirm its efficacy [5]. The results from various studies investigating the use of BoNT as an anti-tumor therapeutic also show promise. In several in vivo experiments, direct injection of BoNT into various malignant tumors demonstrated cellular apoptosis and reduction of tumor size [44–46]. Adding BoNT (Type A) to a diverse range of cancer cell cultures showed slowed cell growth, as well as induction of apoptosis and reduction of mitotic activity [47–52].

Although some cancers have been treated with relative success in the past twenty years, there still remains a paucity of options for patients with difficult to treat, relapsing, or rare cancers. Indeed, cancer is surpassing cardiovascular disease to become the leading cause of death in many populations around the world. This Special Issue presents impactful research that explores the use of toxins as feasible and pertinent cancer therapies which some day may be the solution for so many suffering patients.

Author Contributions: J.K., R.D.C., and A.E.S. conceived the review, J.K. and R.D.C wrote the manuscript, and A.E.S. revised. All authors have read and agreed to the published version of the manuscript.

Funding: The authors are supported, in part, by the Department of Defense Congressionally Directed Medical Research Programs (#W81XWH-17-1-0299, #W81XWH-19-1-0263, and #W81XWH-19-1-0067 to A.E.S.), Targeted Diagnostics and Therapeutics Inc. (A.E.S.), and the DeGregorio Family Foundation (A.E.S.).

Acknowledgments: The Editor is grateful to all the authors who contributed their work to this Special Issue and the exemplary evaluations by the Editorial Board Members and experts who provided peer review for this Special Issue. Finally, I would like to thank the entire MDPI team who helped organize this issue and provided invaluable support throughout the process.

Conflicts of Interest: The authors declare no conflict of interest.

References

1. Poison as Medicine|American Museum of Natural History. Available online: https://www.amnh.org/explore/news-blogs/on-exhibit-posts/the-power-of-poison-poison-as-medicine (accessed on 15 June 2020).
2. Carlson, R.D.; Flickinger, J.C.; Snook, A.E. Talkin' toxins: From coley's to modern cancer immunotherapy. *Toxins* **2020**, *12*, 241. [CrossRef] [PubMed]
3. Debernardi, J.; Pioche-Durieu, C.; Cam, E.L.; Wiels, J.; Robert, A. Verotoxin-1-Induced ER Stress Triggers Apoptotic or Survival Pathways in Burkitt Lymphoma Cells. *Toxins* **2020**, *12*, 316. [CrossRef] [PubMed]
4. Colarusso, A.; Maroccia, Z.; Parrilli, E.; Germinario, E.A.P.; Fortuna, A.; Loizzo, S.; Ricceri, L.; Tutino, M.L.; Fiorentini, C.; Fabbri, A. Cnf1 Variants Endowed with the Ability to Cross the Blood-Brain Barrier: A New Potential Therapeutic Strategy for Glioblastoma. *Toxins* **2020**, *12*, 291. [CrossRef] [PubMed]
5. Mittal, S.O.; Jabbari, B. Botulinum Neurotoxins and Cancer-A Review of the Literature. *Toxins* **2020**, *12*, 32. [CrossRef] [PubMed]
6. Coley, W.B. Contribution to the Knowledge of Sarcoma. *Ann. Surg.* **1891**, *14*, 199–220. [CrossRef] [PubMed]
7. McCarthy, E.F. The toxins of William B. Coley and the treatment of bone and soft-tissue sarcomas. *Iowa Orthop. J.* **2006**, *26*, 154–158.
8. Schreiber, R.D.; Old, L.J.; Smyth, M.J. Cancer immunoediting: Integrating immunity's roles in cancer suppression and promotion. *Science* **2011**, *331*, 1565–1570. [CrossRef]
9. Topalian, S.L.; Hodi, F.S.; Brahmer, J.R.; Gettinger, S.N.; Smith, D.C.; McDermott, D.F.; Powderly, J.D.; Carvajal, R.D.; Sosman, J.A.; Atkins, M.B.; et al. Safety, activity, and immune correlates of anti-PD-1 antibody in cancer. *N. Engl. J. Med.* **2012**, *366*, 2443–2454. [CrossRef]
10. Garon, E.B.; Rizvi, N.A.; Hui, R.; Leighl, N.; Balmanoukian, A.S.; Eder, J.P.; Patnaik, A.; Aggarwal, C.; Gubens, M.; Horn, L.; et al. KEYNOTE-001 Investigators Pembrolizumab for the treatment of non-small-cell lung cancer. *N. Engl. J. Med.* **2015**, *372*, 2018–2028. [CrossRef]
11. Hodi, F.S.; O'Day, S.J.; McDermott, D.F.; Weber, R.W.; Sosman, J.A.; Haanen, J.B.; Gonzalez, R.; Robert, C.; Schadendorf, D.; Hassel, J.C.; et al. Improved survival with ipilimumab in patients with metastatic melanoma. *N. Engl. J. Med.* **2010**, *363*, 711–723. [CrossRef]
12. Brahmer, J.R.; Tykodi, S.S.; Chow, L.Q.M.; Hwu, W.-J.; Topalian, S.L.; Hwu, P.; Drake, C.G.; Camacho, L.H.; Kauh, J.; Odunsi, K.; et al. Safety and activity of anti-PD-L1 antibody in patients with advanced cancer. *N. Engl. J. Med.* **2012**, *366*, 2455–2465. [CrossRef] [PubMed]

13. Howie, J.W. Experiments with interferon in man: A report to the medical research council from the scientific committee on interferon. *Lancet* **1965**, *1*, 505–506. [PubMed]

14. Müller, U.; Steinhoff, U.; Reis, L.F.; Hemmi, S.; Pavlovic, J.; Zinkernagel, R.M.; Aguet, M. Functional role of type I and type II interferons in antiviral defense. *Science* **1994**, *264*, 1918–1921. [CrossRef]

15. Jiang, T.; Shi, T.; Zhang, H.; Hu, J.; Song, Y.; Wei, J.; Ren, S.; Zhou, C. Tumor neoantigens: From basic research to clinical applications. *J. Hematol. Oncol.* **2019**, *12*, 93. [CrossRef]

16. Kimura, T.; McKolanis, J.R.; Dzubinski, L.A.; Islam, K.; Potter, D.M.; Salazar, A.M.; Schoen, R.E.; Finn, O.J. MUC1 vaccine for individuals with advanced adenoma of the colon: A cancer immunoprevention feasibility study. *Cancer Prev. Res.* **2013**, *6*, 18–26. [CrossRef]

17. Lawler, S.E.; Speranza, M.-C.; Cho, C.-F.; Chiocca, E.A. Oncolytic viruses in cancer treatment: A review. *JAMA Oncol.* **2017**, *3*, 841–849. [CrossRef] [PubMed]

18. Andtbacka, R.H.I.; Kaufman, H.L.; Collichio, F.; Amatruda, T.; Senzer, N.; Chesney, J.; Delman, K.A.; Spitler, L.E.; Puzanov, I.; Agarwala, S.S.; et al. Talimogene laherparepvec improves durable response rate in patients with advanced melanoma. *J. Clin. Oncol.* **2015**, *33*, 2780–2788. [CrossRef] [PubMed]

19. Johannes, L. Shiga Toxin-A Model for Glycolipid-Dependent and Lectin-Driven Endocytosis. *Toxins* **2017**, *9*, 340. [CrossRef] [PubMed]

20. Nudelman, E.; Kannagi, R.; Hakomori, S.; Parsons, M.; Lipinski, M.; Wiels, J.; Fellous, M.; Tursz, T. A glycolipid antigen associated with Burkitt lymphoma defined by a monoclonal antibody. *Science* **1983**, *220*, 509–511. [CrossRef] [PubMed]

21. Engedal, N.; Skotland, T.; Torgersen, M.L.; Sandvig, K. Shiga toxin and its use in targeted cancer therapy and imaging. *Microb. Biotechnol.* **2011**, *4*, 32–46. [CrossRef] [PubMed]

22. Fabbri, A.; Travaglione, S.; Rosadi, F.; Ballan, G.; Maroccia, Z.; Giambenedetti, M.; Guidotti, M.; Ødum, N.; Krejsgaard, T.; Fiorentini, C. The Escherichia coli protein toxin cytotoxic necrotizing factor 1 induces epithelial mesenchymal transition. *Cell Microbiol.* **2020**, *22*, e13138. [CrossRef] [PubMed]

23. Vannini, E.; Olimpico, F.; Middei, S.; Ammassari-Teule, M.; de Graaf, E.L.; McDonnell, L.; Schmidt, G.; Fabbri, A.; Fiorentini, C.; Baroncelli, L.; et al. Electrophysiology of glioma: A Rho GTPase-activating protein reduces tumor growth and spares neuron structure and function. *Neuro. Oncol.* **2016**, *18*, 1634–1643. [CrossRef] [PubMed]

24. Frankel, A.E.; Woo, J.-H.; Neville, D.M. Immunotoxins. In *Principles of Cancer Biotherapy*; Oldham, R.K., Dillman, R.O., Eds.; Springer: Dordrecht, The Netherlands, 2009; pp. 407–449.

25. Madhumathi, J.; Verma, R.S. Therapeutic targets and recent advances in protein immunotoxins. *Curr. Opin. Microbiol.* **2012**, *15*, 300–309. [CrossRef] [PubMed]

26. Ruiz-de-la-Herrán, J.; Tomé-Amat, J.; Lázaro-Gorines, R.; Gavilanes, J.G.; Lacadena, J. Inclusion of a Furin Cleavage Site Enhances Antitumor Efficacy against Colorectal Cancer Cells of Ribotoxin α-Sarcin- or RNase T1-Based Immunotoxins. *Toxins* **2019**, *11*, 593. [CrossRef]

27. Tomé-Amat, J.; Ruiz-de-la-Herrán, J.; Martínez-del-Pozo, Á.; Gavilanes, J.G.; Lacadena, J. α-sarcin and RNase T1 based immunoconjugates: The role of intracellular trafficking in cytotoxic efficiency. *FEBS J.* **2015**, *282*, 673–684. [CrossRef]

28. Van Daele, D.J.; Finnegan, E.M.; Rodnitzky, R.L.; Zhen, W.; McCulloch, T.M.; Hoffman, H.T. Head and neck muscle spasm after radiotherapy: Management with botulinum toxin A injection. *Arch. Otolaryngol. Head Neck Surg.* **2002**, *128*, 956–959. [CrossRef]

29. Layeeque, R.; Hochberg, J.; Siegel, E.; Kunkel, K.; Kepple, J.; Henry-Tillman, R.S.; Dunlap, M.; Seibert, J.; Klimberg, V.S. Botulinum toxin infiltration for pain control after mastectomy and expander reconstruction. *Ann. Surg.* **2004**, *240*, 608. [CrossRef]

30. Vasan, C.W.; Liu, W.-C.; Klussmann, J.-P.; Guntinas-Lichius, O. Botulinum toxin type A for the treatment of chronic neck pain after neck dissection. *Head Neck* **2004**, *26*, 39–45. [CrossRef]

31. Wittekindt, C.; Liu, W.-C.; Preuss, S.F.; Guntinas-Lichius, O. Botulinum toxin A for neuropathic pain after neck dissection: A dose-finding study. *Laryngoscope* **2006**, *116*, 1168–1171. [CrossRef]

32. Hartl, D.M.; Cohen, M.; Juliéron, M.; Marandas, P.; Janot, F.; Bourhis, J. Botulinum toxin for radiation-induced facial pain and trismus. *Otolaryngol. Head Neck Surg.* **2008**, *138*, 459–463. [CrossRef]

33. Stubblefield, M.D.; Levine, A.; Custodio, C.M.; Fitzpatrick, T. The role of botulinum toxin type A in the radiation fibrosis syndrome: A preliminary report. *Arch. Phys. Med. Rehabil.* **2008**, *89*, 417–421. [CrossRef] [PubMed]

34. Mittal, S.; Machado, D.G.; Jabbari, B. OnabotulinumtoxinA for treatment of focal cancer pain after surgery and/or radiation. *Pain Med.* **2012**, *13*, 1029–1033. [CrossRef]

35. Bach, C.A.; Wagner, I.; Lachiver, X.; Baujat, B.; Chabolle, F. Botulinum toxin in the treatment of post-radiosurgical neck contracture in head and neck cancer: A novel approach. *Eur. Ann. Otorhinolaryngol. Head Neck Dis.* **2012**, *129*, 6–10. [CrossRef]

36. Rostami, R.; Mittal, S.O.; Radmand, R.; Jabbari, B. Incobotulinum Toxin-A Improves Post-Surgical and Post-Radiation Pain in Cancer Patients. *Toxins* **2016**, *8*, 22. [CrossRef] [PubMed]

37. De Groef, A.; Devoogdt, N.; Van Kampen, M.; De Hertogh, L.; Vergote, M.; Geraerts, I.; Dams, L.; Van der Gucht, E.; Debeer, P. The effectiveness of Botulinum Toxin A for treatment of upper limb impairments and dysfunctions in breast cancer survivors: A randomised controlled trial. *Eur. J. Cancer Care* **2020**, *29*, e13175. [CrossRef] [PubMed]

38. Mailly, M.; Benzakin, S.; Chauvin, A.; Brasnu, D.; Ayache, D. Radiation-induced head and neck pain: Management with botulinum toxin a injections. *Cancer Radiother.* **2019**, *23*, 312–315. [CrossRef]

39. Oh, H.-M.; Chung, M.E. Botulinum toxin for neuropathic pain: A review of the literature. *Toxins* **2015**, *7*, 3127–3154. [CrossRef]

40. Park, J.; Park, H.J. Botulinum toxin for the treatment of neuropathic pain. *Toxins* **2017**, *9*, 260. [CrossRef]

41. Matak, I.; Bölcskei, K.; Bach-Rojecky, L.; Helyes, Z. Mechanisms of botulinum toxin type A action on pain. *Toxins* **2019**, *11*, 459. [CrossRef]

42. Mittal, S.O.; Safarpour, D.; Jabbari, B. Botulinum toxin treatment of neuropathic pain. *Semin. Neurol.* **2016**, *36*, 73–83. [CrossRef]

43. Matak, I.; Lacković, Z. Botulinum neurotoxin type A: Actions beyond SNAP-25? *Toxicology* **2015**, *335*, 79–84. [CrossRef] [PubMed]

44. Vezdrevanis, K. Prostatic carcinoma shrunk after intraprostatic injection of botulinum toxin. *Urol. J.* **2011**, *8*, 239–241. [PubMed]

45. Ulloa, F.; Gonzàlez-Juncà, A.; Meffre, D.; Barrecheguren, P.J.; Martínez-Mármol, R.; Pazos, I.; Olivé, N.; Cotrufo, T.; Seoane, J.; Soriano, E. Blockade of the SNARE protein syntaxin 1 inhibits glioblastoma tumor growth. *PLoS ONE* **2015**, *10*, e0119707. [CrossRef] [PubMed]

46. He, D.; Manzoni, A.; Florentin, D.; Fisher, W.; Ding, Y.; Lee, M.; Ayala, G. Biologic effect of neurogenesis in pancreatic cancer. *Hum. Pathol.* **2016**, *52*, 182–189. [CrossRef] [PubMed]

47. Karsenty, G.; Rocha, J.; Chevalier, S.; Scarlata, E.; Andrieu, C.; Zouanat, F.Z.; Rocchi, P.; Giusiano, S.; Elzayat, E.A.; Corcos, J. Botulinum toxin type A inhibits the growth of LNCaP human prostate cancer cells in vitro and in vivo. *Prostate* **2009**, *69*, 1143–1150. [CrossRef] [PubMed]

48. Nam, H.J.; Kang, J.K.; Chang, J.S.; Lee, M.S.; Nam, S.T.; Jung, H.W.; Kim, S.-K.; Ha, E.-M.; Seok, H.; Son, S.W.; et al. Cells transformed by PLC-gamma 1 overexpression are highly sensitive to clostridium difficile toxin A-induced apoptosis and mitotic inhibition. *J. Microbiol. Biotechnol.* **2012**, *22*, 50–57. [CrossRef]

49. Proietti, S.; Nardicchi, V.; Porena, M.; Giannantoni, A. Botulinum toxin type-A toxin activity on prostate cancer cell lines. *Urologia* **2012**, *79*, 135–141. [CrossRef]

50. Bandala, C.; Perez-Santos, J.L.M.; Lara-Padilla, E.; Delgado Lopez, M.G.; Anaya-Ruiz, M. Effect of botulinum toxin A on proliferation and apoptosis in the T47D breast cancer cell line. *Asian Pac. J. Cancer Prev.* **2013**, *14*, 891–894. [CrossRef]

51. Bandala, C.; Cortés-Algara, A.L.; Mejía-Barradas, C.M.; Ilizaliturri-Flores, I.; Dominguez-Rubio, R.; Bazán-Méndez, C.I.; Floriano-Sánchez, E.; Luna-Arias, J.P.; Anaya-Ruiz, M.; Lara-Padilla, E. Botulinum neurotoxin type A inhibits synaptic vesicle 2 expression in breast cancer cell lines. *Int. J. Clin. Exp. Pathol.* **2015**, *8*, 8411–8418.

52. Rust, A.; Leese, C.; Binz, T.; Davletov, B. Botulinum neurotoxin type C protease induces apoptosis in differentiated human neuroblastoma cells. *Oncotarget* **2016**, *7*, 33220–33228. [CrossRef]

Review

Talkin' Toxins: From Coley's to Modern Cancer Immunotherapy

Robert D. Carlson [†], **John C. Flickinger, Jr.** [†] **and Adam E. Snook** *

Department of Pharmacology and Experimental Therapeutics, Thomas Jefferson University, 1020 Locust Street, Philadelphia, PA 19107, USA; Robert.Carlson@jefferson.edu (R.D.C.); John.Flickinger@jefferson.edu (J.C.F.J.)
* Correspondence: adam.snook@jefferson.edu; Tel.: +1-215-503-7445
† These authors contributed equally to this work.

Received: 11 March 2020; Accepted: 7 April 2020; Published: 9 April 2020

Abstract: The ability of the immune system to precisely target and eliminate aberrant or infected cells has long been studied in the field of infectious diseases. Attempts to define and exploit these potent immunological processes in the fight against cancer has been a longstanding effort dating back over 100 years to when Dr. William Coley purposefully infected cancer patients with a cocktail of heat-killed bacteria to stimulate anti-cancer immune processes. Although the field of cancer immunotherapy has been dotted with skepticism at times, the success of immune checkpoint inhibitors and recent FDA approvals of autologous cell therapies have pivoted immunotherapy to center stage as one of the most promising strategies to treat cancer. This review aims to summarize historic milestones throughout the field of cancer immunotherapy as well as highlight current and promising immunotherapies in development.

Keywords: cancer; immunotherapy; vaccine; immune checkpoint inhibitors; adoptive cell therapy; cytokine therapy; Coley's Toxins

Key Contribution: This review summarizes the pivotal milestones in cancer immunotherapy development from Coley's Toxins to modern day.

---

1. Introduction

The understanding of immune system governance in neoplastic growth and development has made significant leaps in recent years [1], but its origins can be traced back well over a century ago. Incidence of tumors spontaneously regressing following infectious or pyretic periods have been described throughout history [2–4]. However, advancements made in histological diagnosis and assessment of tumor malignancies over the past 100 years have given credence to these claims of immune system modulation in cancer.

2. Pivotal Observations in Cancer Immunotherapy

It is possible that cancer has existed ever since the evolution from unicellular organisms into multicellular entities. However, the oldest record of cancer to date is from a 240 million-year-old fossil containing a shell-less stem turtle, *Pappochelys rosinae*, with evidence of osteosarcoma [5]. Until recently, the treatment of cancer has historically focused on tumor excision, cytotoxic chemotherapeutic agents, and radiation therapy. Only after the turn of the 21st century did immunotherapy to treat cancer take stage [Figure 1].

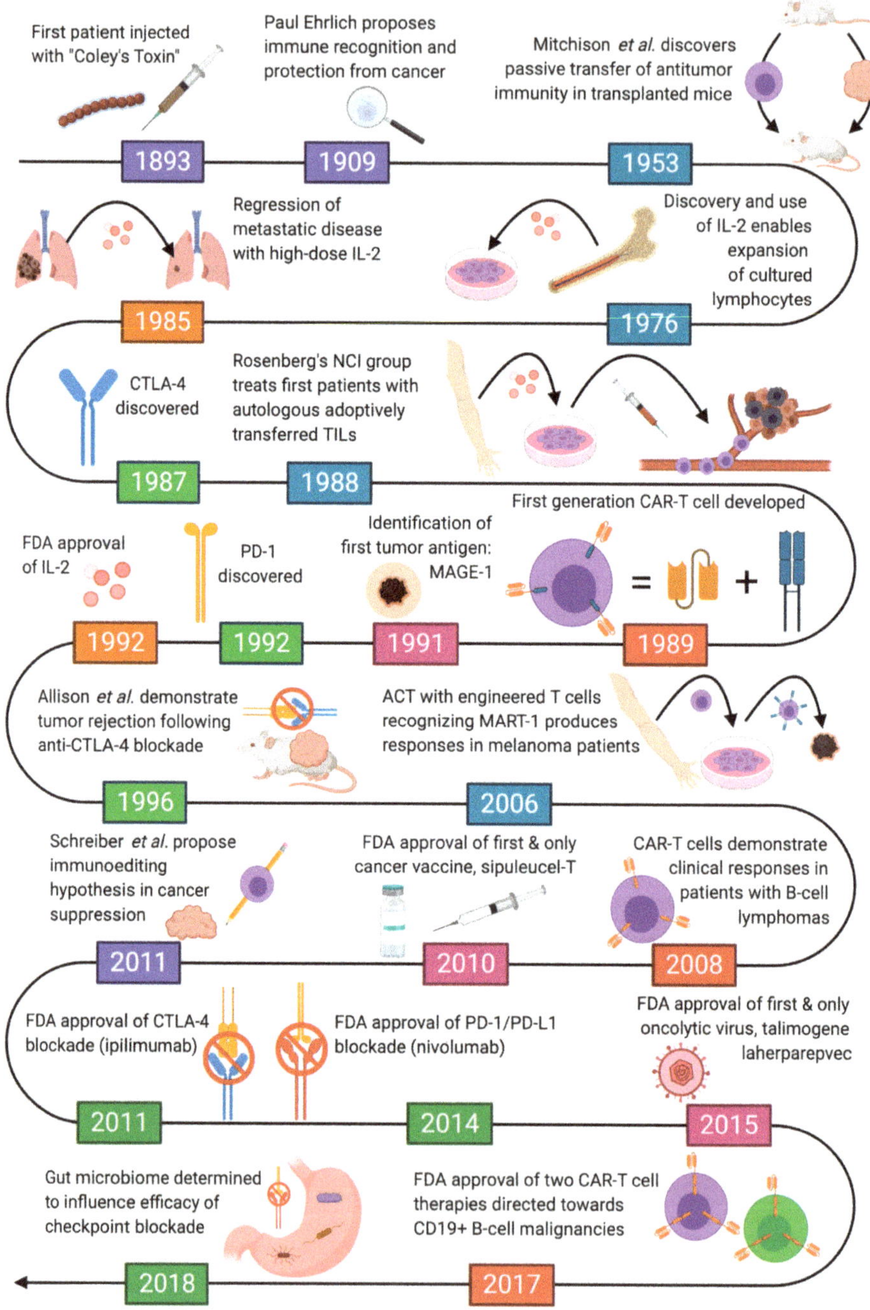

Figure 1. Milestones in the History of Cancer Immunotherapy.

2.1. The Story of Coley's Toxins

William Coley, often regarded as the "Father of Immunotherapy", was a bone surgeon in New York from 1890–1936 who famously developed a cocktail of heat-killed bacteria, called "Coley's Toxins", to treat patients with osteosarcoma. Inspiration for developing this treatment apparently started with one of his first patients, a young woman with osteosarcoma of the hand. Despite his surgical intervention (amputation of the forearm), she succumbed to metastatic disease within months of the operation. This episode had a profound impact on Coley and motivated him to learn more about her disease. He began by reviewing hospital medical records from ninety sarcoma patients, an analysis he later published [6]. While conducting his review, one patient's course of disease was of particular intrigue. Coley came across the description of a patient with an inoperable sarcoma whose tumor completely regressed after developing erysipelas [7], a type of skin infection [8]. Upon reading this account, Coley wondered if it was possible to induce erysipelas in patients as a means to treat cancer. Fortunately for Coley, a German surgeon named Friedrich Fehleisen had only a few years earlier, in 1883, identified *Streptococcus pyogenes* as the bacterium responsible for erysipelas [9]. Thus, Coley was able to test his hypothesis and began injecting sarcoma patients with *Streptococcus pyogenes*, a primitive version of what would later be named Coley's Toxins.

Over the course of Coley's career, from 1888–1933, he tested over a dozen different preparations of his toxin. Developing his infamous toxin required striking a balance between safety and efficacy. Indeed, early preparations were highly variable. Some preparations were impotent and failed to produce any signs of infection while other preparations were highly infectious and led to mortality [10]. Eventually, Coley settled on a combination of heat-killed *Streptococcus pyogenes* and *Serratia marcescens* [11]. Although Coley was not the first person to make a connection between infection and cancer regression, nor the first to inject bacteria into a patient as a means to mediate tumor rejection, Coley's efforts were the most comprehensive and influential. In total, it is estimated that Coley himself injected more than 1000 cancer patients and published over 150 papers related to the topic [11].

Coley reported remarkable success with his toxins and published many reports of his toxins inducing tumor regression [12,13]. However, at the time, his findings were highly controversial and were met with harsh criticism by many of his colleagues. Notable critiques include those in the *Journal of the American Medical Association* in 1894 issuing a statement criticizing the use of his toxins as well as the FDA re-categorizing of "Coley's Toxins" in 1963 as an investigational drug that lacked safety and efficacy data, despite over 70 years of use and numerous publications [11]. This recategorization made it illegal to prescribe Coley's Toxins outside of clinical trial testing. In the end, history would be on the side of William Coley. Years after his death, his toxins were re-evaluated in a controlled trial and were demonstrated to mediate antitumor effects [14]. Moreover, advancements in fundamental understanding of cancer and the immune system have allowed his findings to become more widely accepted and to lay a foundation for future studies of cancer immunotherapy.

2.2. Evidence the Immune System Targets Cancer

Although Coley never fully understood the mechanism by which his toxins functioned, he gathered substantial evidence linking the immune system and cancer. Further clarity and development of this connection would come years later in the form of the immunosurveillance hypothesis. The idea that the immune system possesses a capacity to recognize and eliminate cancer cells was first postulated by Paul Ehrlich in 1909 [15]. While direct experimental evidence during this time period was lacking, Ehrlich reasoned that the incidence of cancer is relatively low but that the formation of aberrant cells is a common phenomenon, suggesting the existence of a host defense system against cancer. Over 50 years later, these ideas were further developed by Burnet and Thomas and formally coined the "immune surveillance" hypothesis [16,17].

Early experimental evidence for the existence of tumor-specific immunity derives from transplantation studies. In 1943, Luwik Gross utilized methylcholanthrene (MCA) to chemically induce sarcoma in a C3H mouse and then transplanted this sarcoma into syngeneic mice. While inoculation with high doses of tumor cells often killed mice, Gross found that inoculation with low doses of tumor cells led to a period of growth followed by gradual tumor regression. In these surviving mice, tumor challenge using high doses of tumor cells invariably led to rejection, suggesting these animals developed immunity to the tumor [18]. Further support for immunosurveillance comes from a seminal study by Prehn and Main in 1953. In these studies, an array of sarcomas from multiple syngeneic mice were generated using MCA. Prehn and Main found that inoculation of a mouse with sarcoma from one source protected that mouse from future challenge using the same sarcoma source but did not protect against challenge using sarcoma derived from a different mouse [19]. Moreover, Prehn and Main demonstrated that transplantation of skin tissue from a donor mouse did not sensitize the recipient mouse to the donor's sarcoma, directly addressing a common critique at the time that rejection was mediated by subtle differences in genetic backgrounds. Collectively, these studies further supported the existence of tumor-specific immunity, adding the nuance that tumor antigens are highly unique to a tumor even in tumors of the same histological type, induced by the same chemical means, and from mice of the same genetic background [19].

While studies in partially immunocompromised mouse models over the following decades failed to support the immunosurveillance hypothesis, definitive demonstration of immunosurveillance came in the early 2000s following a series of studies conducted in novel, specifically immunocompromised, mouse strains. In 2001, Robert Schreiber's group compared the incidence of spontaneous neoplasms between wild-type and $Rag2^{-/-}$ mice ($Rag2$ encodes a protein necessary for somatic recombination and thus $Rag2^{-/-}$ mice lack mature T and B lymphocytes) [20]. In mice over 15 months old, fewer than 20% of wild-type mice contained neoplastic disease while 100% of surveyed $Rag2^{-/-}$ mice developed spontaneous neoplastic lesions in various tissues, suggesting functional T and B lymphocytes suppress the development of cancer. Moreover, the same study observed that $Rag2^{-/-}$ mice, as well as $Ifngr1^{-/-}$ and $Stat1^{-/-}$ mice, which are deficient in vital immune signaling pathways, develop higher incidences of sarcoma compared to wild-type mice in MCA-induced tumor models [21]. Similarly, higher incidences of MCA-induced tumors were reported by additional investigators using mice deficient in other vital immune-signaling molecules such as perforin or TNF-related apoptosis-inducing ligand (TRAIL) [22,23]. These experimental studies in mice are mirrored by clinical evidence that humans with compromised immune systems develop higher incidences of cancer. Indeed, individuals born with genetic defects in immune-related genes develop higher incidences of lymphoma [24]. Moreover, people with otherwise normal immune function who acquire AIDS infection or transplant patients who receive immunosuppressive drugs are both at higher risk for developing Non-Hodgkin's lymphoma and virus-induced Kaposi Sarcoma [25,26].

Since the emergence of the immunosurveillance hypothesis, the interplay between the immune system and cancer has been further refined and renamed as the process of "immunoediting" [27]. Immunoediting posits that the immune system and cancer intersect at three stages: elimination, equilibrium, and escape. During the elimination stage, the immune system recognizes and destroys many, but not all aberrant cells. During equilibrium, the immune system and the tumor exert opposing forces, effectively resulting in containment of the tumor. Over time, as the cancer acquires additional mutations and as the immune system exerts a selective pressure eliminating immunogenic cells and leaving behind non-immunogenic cells, the cancer eventually fully escapes immune surveillance. In this final escape stage, the cancer has fully circumvented detection by the immune system and undergoes rapid and uncontrolled growth. Evidence for equilibrium and escape stages is supported by experiments in mice with stagnant tumor sizes who then undergo rapid growth after immune cell depletion [28]. Additionally, tumors arising from immunocompromised mice are more frequently rejected when transplanted into a wild-type host compared to tumors arising from immunocompetent mice [21], partly reflecting immune-induced antigen loss in the presence of an intact immune system [29] (immunoediting). Thus, the fundamental goal of cancer immunotherapy is to overcome the years to decades of immunoediting to generate antitumor immunity that is sufficient to completely eliminate the patient's cancer and cure their disease.

3. Immunomodulatory Agents

Immunomodulators comprise a variety of therapeutic agents and treatment strategies that aid in normalizing, or in the context of cancer, re-engaging or boosting immune cell function to counter uncontrolled cell proliferation. By definition, a tumor that presents in the clinic can be said to have escaped normal immune control, even if tumor-reactive T cells are detected in the blood or have infiltrated the tumor tissue [30]. Although tumor cells themselves possess intrinsic immunosuppressive behaviors, such as cultivating a hypoxic microenvironment [31,32] or generating lactic acid [33], the majority of suppressive influence comes from the normal functions of suppressive immune cells, cytokines, and inhibitory surface molecules [34,35]. Under normal physiological circumstances, these mechanisms suppress T-cell priming and cytotoxic T-cell function to stave off unwarranted autoinflammatory responses [36]. Therefore, many of the immunomodulatory agents described herein aim to directly oppose these immunosuppressive mechanisms and to re-engage the immune system.

3.1. Cytokine Therapy

Prior to understanding their therapeutic immunomodulatory potential in cancer, cytokines were first recognized as systemic soluble factors that could regulate lymphocyte function and inflammatory responses. In 1972, a group from Yale University School of Medicine first characterized a "lymphocyte activating factor" that spurred lymphocyte proliferation in response to soluble agents released by other syngeneic immune cells [37]. These agents were partially purified from antigen-stimulated, lymphocyte-conditioned media and characterized further as "T-cell growth factor" that could support cytotoxic T cells capable of killing autologous leukemic myeloblasts [38]. Shortly thereafter, this key growth factor was definitively purified and described as interleukin-2 (IL-2) [39], which not only allowed T lymphocytes to be cultured in large quantities ex vivo, but also allowed recombinant IL-2 to be administered as a high-dose single-agent [40], or used in tandem with preconditioned and transplanted cancer-specific lymphoid cells [41,42]. A comprehensive report published in 1987 by the NCI's Surgery Branch documented objective responses to high-dose IL-2 and regression of tumors in patients with metastatic melanoma and renal cell cancer [43]. Adjustments made to IL-2 dose scheduling would largely combat acute toxicities, the most prominent being capillary leak syndrome and hypovolemia [44]. These results and safety measures would spur numerous and larger cohort studies utilizing IL-2 in a metastatic setting [45,46], culminating with FDA approval of high-dose intravenous IL-2 for patients with metastatic renal cancer in 1992, and metastatic melanoma several years later. This would be noted as one of the first FDA-approved cancer immunotherapies [47].

Other cytokines that have demonstrated translational applications include interleukin-15 (IL-15), interferon-alpha (IFNα), and granulocyte macrophage colony-stimulating factor (GM-CSF). While both structurally similar and capable of stimulating early T-cell proliferation and NK cell engagement much like IL-2, IL-15 additionally supports memory CD8$^+$ T-cell persistence, a known mediator of long-term antitumor immunity [48]. IL-15 has also proven to be an effective mediator of antitumor protection in murine models of cancer [49–52]. In support of these claims, phase 1 clinical trials utilizing recombinant IL 15 alone, and in conjunction with B-cell-depleting antibodies, are currently underway for treating both solid and liquid tumors, respectively [NCT01021059, NCT03759184].

IFNα, another cytokine originally described in the context of mediating antiviral immune response [53,54], was also identified to inhibit tumor neovasculature, upregulate MHC class I expression, mediate dendritic cell maturation, activate B and T cells, and induce apoptosis—all favorable antitumor attributes [55]. Thanks to the efforts of blood banking that began in the late 1970s [56], adequate quantities of purified IFNα spurred a burst of clinical evaluations in patients with hematological malignancies [57,58] and solid tumors, such as renal cell cancers and malignant melanoma [59,60]. These trials culminated in FDA approval of IFNα as an adjuvant therapy first in rare forms of leukemia, and later in patients with high-risk stage II and stage III melanoma [61].

As the name suggests, GM-CSF was originally identified as a regulator of granulocyte and macrophage differentiation, as well as general hematopoiesis of multi-lineage progenitors [62]. However, in 1993,

Dranoff and colleagues transduced B16 melanoma cells with ten known pro-inflammatory mediators, vaccinated mice with these constructs, and then challenged them with live B16 cells. Of the ten, GM-CSF conferred the largest magnitude of antitumor immunity [63]. These findings prompted strategies to deliver GM-CSF to patients, either by vaccinating patients with irradiated tumor cells engineered to secrete the cytokine [64], or by single-agent dosing [65]. Although the exploration of GM-CSF-expressing tumor vaccines has waned in recent years due to limited clinical efficacy [66], combination strategies employing recombinant GM-CSF with other immunomodulatory agents, such as checkpoint inhibitors [67] and additional cytokines [68] have enhanced overall survival in melanoma patient trials.

3.2. Immune Checkpoint Inhibitors

Suppression of immune cell activation, infiltration, and effector functions required for tumor cell clearance can be largely attributed to the immunosuppressive conditions within the tumor microenvironment [69]. In several studies, "dysfunctional" CD8$^+$ T cells were retrieved from patient tumors and nearby draining lymph nodes and said to be lacking the expected differentiation profiles [70], or impaired by the accumulation of repressive Foxp3$^+$ regulatory T cells (Tregs) [71]. A similar phenomenon was originally described in mice infected with a lymphocytic choriomeningitis virus, whereby chronic antigen stimulation induced an "exhausted" T-cell state proceeded by T-cell receptor (TCR) downregulation [72]. The discovery of other requisite activating co-stimulatory signals in addition to canonical antigen stimulation with the TCR [73], which was now known to be insufficient for fully functional T-cell activation [74], gave clues to the complex nature of balancing activation with self-antigen tolerance [75]. Immune homeostasis is dependent upon these co-stimulatory/co-inhibitory receptor-ligand interactions, which in the correct context, safeguard against chronic immune activation and excessive inflammatory responses [76]. Addressing these phenomena directly, immune checkpoint blockade selectively restricts these co-inhibitory signaling mechanisms that have been co-opted by cancer cells, thereby enhancing antitumor T-cell activity.

Although initially identified in 1987 [77], the 1994 discovery of cytotoxic T lymphocyte-associated antigen-4 (CTLA-4) co-inhibitory receptor pairing with the B7 co-stimulatory ligand is perhaps the most substantial [78]. Upregulation of CTLA-4 on both CD4$^+$ and CD8$^+$ T lymphocytes was identified as a negative regulator of T-cell activation and effector functions [79,80], while murine models deficient in CTLA-4 experienced massive lymphoproliferation and tissue infiltration due to over-activation of resident T cells [81,82]. In the late 1990s, Dr. James Allison's group at University of California, administered an inhibitory antibody to block the CTLA-4 co-inhibitory synapse in mice burdened with tumors. Both orthotopic and pre-established tumor cells were rejected following administration of the anti-CTLA-4 antibody, indicating that blockade of inhibitory signals associated with the co-stimulatory pathway can enhance antitumor immunity [83]. These indications prompted the application of CTLA-4 blockade in patients with stage III/IV unresectable melanoma with remarkable success [84], culminating in the 2011 FDA approval of the anti-CTLA-4 monoclonal antibody (mAb), ipilimumab, as an adjuvant therapy for patients with cutaneous melanoma. Retrospective studies have revealed marked increases in survival benefit compared to traditional chemotherapy regimens [85,86], with modest gains observed in other solid tumor types currently in various phases of clinical trials [87].

In an effort to identify genes associated with apoptosis, Dr. Tasuku Honjo's group at Kyoto University discovered programmed cell death protein 1 (PD-1), a lymphoid cell surface protein that the group hypothesized to be a cell-death inducer [88]. Several years later in 1999, the same group generated a PD-1 deficient mouse model that spontaneously developed several autoimmune-like symptoms and systemic graft-versus-host-like disease [89]. Like CTLA-4, PD-1 was identified as a negative regulator of adaptive immune responses. PD-1 ligand 1 (PD-L1) was discovered that same year at the Mayo Clinic and characterized as functionally homologous to the CTLA-4 ligand, B7, but co-stimulated T cells through some additional unknown receptor [90], later identified to be PD-1 [91]. Engagement of PD-1 with its ligand prevented T-cell proliferation and cytokine production when synthetically stimulated, identifying it as an intrinsic inhibitory mechanism of autoreactive lymphocyte activation [91]. PD-L1 surface expression on tumor cells was also discovered to suppress the cytolytic effector functions of CD8$^+$ T cells, with additional speculation that PD-1/PD-L1 blockade could serve as an effective strategy to combat tumor cell escape [92]. Speculation became reality when several groups tested PD-L1 blockade in murine tumor models and concluded that antibodies directed at this co-stimulatory interaction could enhance cancer immunotherapy [93–95]. In one step closer to the clinic, PD-L1 was determined to be a prognostic marker of patient outcome, with higher levels of ligand in resected specimens correlated with poorer patient survival [96]. Within the past 10 years, several high-profile trials employing anti-PD-1/PD-L1 mAbs under various conditions, dosing strategies, and cancer types, have indicated that blockade of this co-inhibitory pathway is both well-tolerated and associated with durable objective responses in patients [97–99]. Consequently, FDA approval was granted first to nivolumab, a humanized PD-1 mAb for metastatic melanoma in 2014, and subsequently for pembrolizumab, a PD-1 mAb alternative. Both therapies elicited greater overall patient survival compared to their anti-CTLA-4 counterpart [100,101]. In 2016, a third antibody was developed, this time directed at the PD-L1 ligand to treat patients with urothelial carcinoma and non-small cell lung cancer with much success [102,103]. This fully humanized anti-PD-L1 mAb, atezolizumab, was granted FDA approval for bladder cancer patients ineligible for traditional cisplatin-based chemotherapies [104]. Recent studies have expanded the number of indications for anti-PD-1/PD-L1 blockade alone [105], and in combination with anti-CTLA-4 [106], both proving to be summarily efficacious.

Although CTLA-4 and PD-1 blockade strategies have demonstrated unprecedented clinical success and accelerated FDA approval, there remains a population of non-responders. These individuals either fail to respond to checkpoint blockade from treatment onset due to innate resistance mechanisms, or acquire secondary resistance resulting in relapse. Retrospective studies of large-cohort clinical trials may expose novel biomarkers capable of predicting resistance to checkpoint therapies [107]. Additional co-inhibitory receptors, each with unique functions, have since been identified to influence negative immune regulation by various mechanisms [108]. Likewise, recent findings have demonstrated that the resident gut microbiome has the ability to influence patient responses to checkpoint blockade, with individuals who had consumed oral antibiotics prior to therapy experiencing poorer anti-PD-1 responses [109,110]. Nonetheless, immune checkpoint inhibitors continue to represent the vast majority of new immunotherapies for the treatment of cancer. These therapies would not be possible without seminal discoveries made in the blockade of negative immune regulatory elements by Drs. James Allison and Tasuku Honjo, for which they were awarded the 2018 Nobel Prize in Physiology or Medicine.

4. Vaccines

While immunomodulatory agents broadly stimulate the immune system, cancer vaccines aim to more precisely steer an immune response towards cancer. At its purest form, a cancer vaccine consists of one or more tumor antigens combined with an adjuvant to enhance the immune response. As will later be described, the type of tumor antigen, delivery method of the antigen, and adjuvant varies greatly. Cancer vaccines can be administered as a therapeutic vaccine in patients with existing malignancies or as a preventive vaccine in healthy or high-risk individuals (primary prevention) or patients in remission (secondary prevention) to protect against future tumor development or recurrence, respectively.

4.1. Tumor Antigens

Initial attempts at vaccination occurred before the identification of specific tumor antigens. These trials utilized cellular-based vaccines consisting of modified or irradiated tumor cells derived from a patient (autologous) or from a cancer cell line (allogeneic) injected with adjuvant [111–113]. With the identification of the first human tumor antigen, MAGE-1, by Thierry Boon's group in 1991, a more refined approach of vaccinating against specific targets was born [114]. Since then, over 75 tumor-associated antigens have been identified [115]. There are two categories of tumor antigens: tumor-associated antigens and tumor-specific antigens. By definition, tumor-associated antigens share expression with some normal tissues while tumor-specific antigens are unique to cancer cells. Notable examples of tumor-associated antigens that have been a focus of multiple immunotherapies include the cancer-testis antigens NY-ESO-1 [116] and MAGE-1 [117], which are expressed by germ cells and ectopically re-expressed in cancers; the oncofetal antigens CEA [118] and alpha-1-fetoprotein [119], which are present during fetal development and re-expressed by cancers; differentiation antigens, such as prostate-specific antigen (PSA) [120] and CD19 [121], which are expressed by cells derived from a specific tissue-type and retained in cancers; and antigens that are over-expressed by cancers relative to normal tissue, including HER2 [122] and telomerase [123]. In contrast to tumor-associated antigens, which share expression with healthy tissue, tumor-specific antigens are exclusively expressed by tumors. Tumor-specific antigens, also known as neoantigens, are mutated peptides created by unique genetic aberrations or may be viral antigens in the case of virus-associated cancers [124].

4.2. Therapeutic Cancer Vaccines

In 1995 and 1996, the first clinical trials testing cancer vaccines against tumor-associated antigens were published. These trials utilized either peptide vaccines [125], or peptide-pulsed dendritic-cell vaccines composed of patient-derived dendritic cells that have been incubated with a peptide prior to re-infusion [126,127]. Other popular methods of cancer vaccination include the use of recombinant viral and bacterial vectors. As microorganisms potently stimulate the immune system, the use of these vectors to deliver tumor antigen in the context of an infection is hypothesized to enhance antitumor immune responses. Common vectors used to deliver tumor antigens include poxvirus [128], adenovirus [129], *Salmonella typhimurium* [130], and *Listeria monocytogenes* [131]. As methods of gene therapy have advanced over the years, the use of DNA and RNA vaccines has become increasingly common [132].

Despite thousands of cancer vaccine clinical trials, only one therapeutic cancer vaccine, Sipuleucel-T, is FDA-approved [133]. Sipuleucel-T is an autologous cell vaccine composed of patient peripheral blood mononuclear cells (PBMCs) pulsed with a chimeric protein consisting of the tumor-associated antigen prostate alkaline phosphatase (PAP) fused to the immunomodulating cytokine GM-CSF. A phase III clinical trial in men with metastatic castration-resistant prostate cancer found that three infusions of Sipuleucel-T led to the induction of PAP-specific immune responses and a 4.1-month improvement in median survival [134].

Limited success in therapeutic settings may be, in part, attributed to poor immunogenicity of the vaccine target and immunosuppressive tumor microenvironments. One approach to overcome poorly immunogenic tumor-associated antigens is the recent trend towards targeting neoantigens. Compared to tumor-associated antigens, neoantigens may be more immunogenic due to a lack of immunological tolerance mechanisms [135]. Until recently, the identification of neoantigens was impractical as the cost and time to sequence patient genomes for unique mutations presented a formidable barrier. However, with advancements in next-generation sequencing, it has become feasible to sequence a patient's normal and tumor genome to identify unique tumor-specific antigens. Personalized therapeutic neoantigen vaccines have shown promise in phase I trials for melanoma [136] and glioblastoma [137]. However, these neoantigen vaccines are in early clinical testing, and thus the efficacy and feasibility of this approach is yet to be determined.

4.3. Preventive Cancer Vaccines

Recently, there has been a trend towards testing cancer vaccines as preventive therapies. Vaccination in preventive settings may be preferable to therapeutic ones as it may allow for the induction of antitumor immunity before the development of immunosuppressive microenvironments [138]. This strategy has shown promise against multiple viral-based cancers. Indeed, vaccination against oncogenic viruses including hepatitis B and human papillomavirus have led to reductions in hepatic [139] and cervical [140] cancers, respectively. However, for non-virally associated cancers, a target antigen and clinical setting to administer preventive vaccines is often less clear. For example, vaccinating a healthy patient against a tumor-associated antigen may carry an unnecessary risk of autoimmunity. Additionally, preventive vaccination against neoantigens, while reducing the risk of autoimmunity, may be impractical as neoantigens are often widely variable between patients. However, preventive vaccination in some settings may be possible. One such example is vaccination against the mucin 1 (MUC1) antigen in patients at high-risk of colorectal cancer. In tumors, MUC1 is hypoglycosylated relative to normal tissues, allowing for the induction of selective antitumor responses [141]. A phase I/II study in patients with a history of colorectal adenoma demonstrated MUC1 immunogenicity and a phase II trial investigating the ability of MUC1 vaccine to prevent adenoma recurrence is currently ongoing [142].

4.4. Oncolytic Virotherapy

An emerging immunotherapeutic strategy that is often categorized as a cancer vaccine is the use of oncolytic viruses. Oncolytic viruses preferentially infect and kill tumor cells compared to normal tissue. Selective infection of tumor cells is achieved through a combination of factors including the overexpression of viral receptors on tumor cells which can facilitate viral entry, a proliferative cell cycle that promotes viral replication, and a tumor cell deficiency in type I interferon production leading to limited viral clearance [143]. In addition to mediating tumor regression by direct cell lysis, viral infection activates components of the innate and adaptive immune system, thereby contributing further to tumor cell death. For example, oncolytic viral infection activates NK cells to clear virally-infected tumor cells [144]. Moreover, immunogenic cell death of virally infected tumor cells releases both tumor-associated antigens and neoantigens that can be acquired and presented by antigen-presenting cells, leading to the induction of antitumor CD8$^+$ T cell responses (an approach often described as "in situ vaccination") [145,146].

The potential of oncolytic virotherapy was first noted by George Dock in 1904. Similar to Coley, Dock noticed that a patient with acute leukemia underwent remission after acquiring an influenza infection [147]. Many other case reports followed over the years, eventually leading to hundreds of clinical trials testing oncolytic viruses [148]. In 2015, the first oncolytic viral therapy, talimogene laherparepvec ("T-VEC"), was approved by the FDA for use in metastatic melanoma [143,149]. T-VEC is an attenuated herpes simplex virus harboring various genetic deletions and insertions designed to enhance the antitumor immune response, such as the deletion of an immune-evasive viral gene *ICP47* and the insertion of a human GM-CSF gene [145]. Compared to GM-CSF administration alone, T-VEC led to a 4.4 month increase in median survival in a phase III trial in patients with advanced and metastatic melanoma [143,149].

5. Adoptive Cell Therapy

5.1. Tumor-Infiltrating Lymphocytes and Engineered T-Cell Receptors

The antitumor activity of T lymphocytes was first elucidated in the 1950s and 1960s with seminal discoveries made in allograft rejection of sarcomas in experimental rodent models [150,151]. In 1953, Mitchison investigated the passive transfer of tumor immunity via transplantation of lymph nodes from mice inoculated with lymphosarcomas to equivalently challenged, but non-inoculated, mice [152]. A decade later, two groups made similar observations of transferrable tumor immunity by isolating and transplanting the cells of lymphatic tissues, rather than the organs themselves. Cells were collected from the lymphatic ducts and spleens of donor animals previously immunized with sarcoma cells that developed palpable tumors. Administration of those lymphoid cells back into syngeneic and non-syngeneic recipients inoculated with tumors, saw sustained regression indicating that these lymphocytes were "educated" by prior exposure to tumor antigens [153,154].

The means to exploit these T lymphocytes for their antitumor potential was limited by the inherent difficulty of expanding cells ex vivo. In 1976, a group at the NIH first described the co-culture of isolated human bone marrow with conditioned media containing IL-2 that could induce growth and differentiation of bone marrow cells to T lymphocyte precursors [155]. With the advent of commercially synthesized IL-2, T lymphocytes could now be cultured in large quantities, or in the context of an adjuvant, to boost the therapeutic effects of tumor-sensitized and adoptively transferred T lymphocytes [156]. This subset of cytokine-activated lymphocytes was identified to be among those infiltrating the stroma of transplanted tumors. Dr. Steven Rosenberg's group confirmed that tumor-infiltrating lymphocytes (TILs) isolated from resected tumor could recognize syngeneic tumor cells in vitro [157], as well as mediate durable antitumor responses when re-administered back into autologous animal models [158] and cancer patients with metastatic disease [159].

The relatively pure populations of CD8$^+$ and CD4$^+$ T cells cultured from resected tumors appeared to dissipate quickly when returned to patients, meaning that therapeutic responses were often transient. However, in 2002, prior application of a lymphodepleting, nonmyeloablative chemotherapy regimen, originally designed for allogeneic bone marrow transplantation [160], greatly enhanced TIL engraftment and clonal persistence in patients, with some individuals harboring up to 80% CD8$^+$ T cells many months post-infusion [161,162]. Shortly thereafter, another milestone was achieved when the Rosenberg group retrovirally transduced patient-derived peripheral blood lymphocytes with a TCR recognizing the common melanoma antigen, MART-1. Objective cancer regression was observed in 2 out of 15 patients (13%) when engineered T lymphocytes were adoptively transferred back into patients [163], with a subsequent report demonstrating an improved 30% objective response rate [164]. Additional trials employing engineered TCRs targeting NY-ESO-1 in synovial cell sarcoma [165], the GD2 disialoganglioside in neuroblastoma [166], and carcinoembryonic antigen (CEA) in colorectal cancer [167] demonstrated objective clinical responses, thereby broadening the application to additional tumor types. More recently, personalized strategies using whole-exome sequencing of patient tumors has given researchers the ability to target neoantigens with high specificity [168,169]. Classically unmanipulated TIL therapy will continue to serve a patient population with shared and broadly targetable antigens [170], while more nuanced TILs recognizing neoantigens will continue to pace evolving therapies in the age of personalized medicine [171].

5.2. Chimeric Antigen Receptor T Cell (CAR-T Cell) Therapy

Although adoptive transfer of tumor-sensitized and antigen-reactive TILs with prefatory lymphodepletion and IL-2 dosing regimens had proven effective in the clinic, patient responses were often transient: shrinkage in metastatic lesions could occur, without objective response to treatment [172]. Native TCRs are often limited by their ability to recognize post-translationally or aberrantly modified proteins, such as those observed in tumor-associated antigens of malignant cells [173,174]. Likewise, T cells dependent on antigen presentation by MHC molecules are routinely hindered by MHC class I downregulation, a selective mechanism of tumor immune escape [175].

In 1989, an Israeli group at the Weizmann Institute of Science devised the first proof-of-concept strategy using an engineered chimeric antigen receptor (CAR) to circumvent the need for MHC-mediated antigen presentation for T-cell activation. By combining the intracellular T-cell receptor circuitry with the antigen-recognizing variable domains of an antibody raised against 2,4,6-trinitrophenyl (TNP), the researchers were able to elicit a non-MHC-restricted response in transfected T lymphocytes to target cells bearing TNP on their surface [176]. The unprecedented antibody-type specificity, now liberated from MHC presentation and paired to effector T-cell function, could conceivably target post-translationally modified proteins characteristic of tumor cells undergoing selection or escape. Several years later, the same group successfully generated CARs directed towards HER2, a cell surface antigen commonly overexpressed in adenocarcinomas. These CAR-T cells selectively lysed HER2$^+$ cancer cells in vitro, providing evidence that mAbs directed towards common tumor cell antigens, could be reassembled into single chain variable fragments (scFvs) to facilitate immune effector function directly [177]. That same year, a joint venture between Weizmann Institute and the NIH expanded the spectrum of available targets by targeting folate receptors commonly overexpressed in ovarian carcinoma cells and further demonstrating the potential of adoptively transferring these CAR-T cells into cancer patients [178].

In an effort to improve CAR-T cell activation, CD28 costimulatory molecules were added in a single tandem gene product with the intracellular CD3ζ-chain. Tumor cells often lack costimulatory molecules entirely, a barrier to persistent activation in first-generation CAR-T cells. Much like conventional T cells, an "exhausted" phenotype was observed in T cells expressing first-generation constructs encountering tumor cells in excess. In contrast, second-generation CAR-T cells containing additional built-in CD28 costimulatory moieties demonstrated superior signaling functionality, persistence, and cytokine production [179,180], as well as antitumor activity [181]. Over the next decade, second-generation CARs would be the basis for many first-in-human studies: first targeting carbonic anhydrase IX (CAIX), an antigen commonly overexpressed in renal cell carcinoma (RCC), and shortly thereafter, the ovarian cancer–associated antigen α-folate receptor (FR). CAIX-directed CAR-T cells produced grade 2–4 liver toxicity in patients due to CAIX self-antigen present in normal bile duct epithelium, with no overall response in RCC tumors [182]. Likewise, FR-directed CAR-T cells in a parallel phase 1 study, produced no reduction in ovarian tumor burden, albeit with lower grade 1–2 toxicity and no detectable off-tumor or off-target responses [183].

Major clinical breakthroughs were not seen until several years later when CAR-T cell therapy strategy switched from targeting primarily solid tumors, to liquid tumors, such as B-cell lymphomas and leukemias. In 2008, a group at University of Washington pioneered a proof-of-concept clinical trial in which refractory B-cell lymphoma patients received 20 infusions of autologous CD20-directed CAR-T cells. Treatment was well-tolerated in patients, with minimal toxicities and modified T cells persisting up to 9 weeks. Of the seven patients, two were noted as having complete response to treatment [184]. With phase 1 clinical trials rapidly taking shape around B-cell targets, methods for manufacturing and validating clinical-grade autologous CAR-T cells were developed to support increasing demand [185]. Shortly thereafter, the Rosenberg group within the NCI's Surgery Branch, demonstrated in vivo antigen-specific activity of CAR-T cells directed towards the B-cell-specific antigen, CD19, in advanced-stage follicular lymphoma [186]. Paralleling this seminal study, Dr. Carl June's group at the University of Pennsylvania demonstrated specific and effective on-target killing of CD19[+] malignant B cells in patients with advanced chronic lymphocytic leukemia (CLL) using CD19-directed CAR-T cells. In that study, two out of three patients experienced complete remission of disease, with a portion of CAR-T cells retaining potent effector function six months after initial infusion, indicating a possible memory CAR-T cell phenotype [187,188]. The CAR employed by the University of Pennsylvania possessed a 4-1BB (CD137) costimulatory domain, rather than CD28, that promoted in vivo persistence and antileukemic function that outperformed conventional CARs with either CD3ζ and CD28 costimulatory molecules or CD3ζ alone [187,189].

Over the next few years, both groups continued to advance the field by targeting various CD19[+] hematological malignancies with great success. However, unanticipated and oftentimes severe neurological toxicities in the form of cytokine release syndrome were observed in patients. This toxicity can manifest as fevers, headaches, aphasia, and in some cases, hallucinations, delirium, and seizures [190]. Cytokine blockade strategies to control the abundance of systemically released cytokines, namely administering the IL-6-blocking antibody tocilizumab with and without corticosteroids, were developed to combat acute neural toxicity [191–195]. Although major clinical gains have been achieved with CD19[+] hematologic malignancies, the same successes have yet to be fully realized in solid tumors. Despite abundant antigenic heterogeneity, difficulties in trafficking to tumor sites, and an intrinsic immunosuppressive tumor microenvironment [196,197], CAR-T cell therapies against solid tumor malignancies have entered early-phase clinical trials with varying degrees of success [Table 1].

Table 1. Summary of active clinical trials for CAR-T cell therapy in solid tumors.

Antigen Target	Cancer Type	Phase	ClinicalTrials.gov Designation
CD117	Osteoid Sarcoma, Ewing Sarcoma	I/II	NCT03356782
CD133	Liver, Pancreatic, Brain, Breas, Ovarian, Colorectal, Acute Myeloid and Lymphoid Leukemias	I/II	NCT02541370
	Osteoid Sarcoma, Ewing Sarcoma	I/II	NCT03356782
CD171	Neuroblastoma, Ganglioneuroblastoma,	I	NCT02311621
CEA	Colorectal	I/II	NCT02959151
	Lung, Colorectal, Gastric, Breast, Pancreatic	I	NCT02349724
EGFR	Colorectal	I/II	NCT03152435
EGFRvIII	Glioma, Glioblastoma, Gliosarcoma	I/II	NCT01454596
EpCAM	Colon, Esophageal, Pancreatic, Prostate, Gastric, Hepatic	I/II	NCT03013712
EphA2	Glioma	I/II	NCT02575261
ErbB	Head and Neck	I/II	NCT01818323
FRα	Urothelial Bladder	I/II	NCT03185468
GD2	Glioma	I/II	NCT03252171
	Neuroblastoma	I/II	NCT03373097
	Neuroblastoma	I	NCT01822652
	Neuroblastoma	II	NCT02765243
	Cervical	I/II	NCT03356795
	Osteoid Sarcoma, Ewing Sarcoma	I/II	NCT03356782
	Sarcoma, Osteosarcoma, Neuroblastoma, Melanoma	I	NCT02107963
GPC3	Hepatocellular Carcinoma	I/II	NCT02723942
	Hepatocellular Carcinoma	I/II	NCT02959151
	Hepatocellular Carcinoma	I/II	NCT03084380
HER2	Breast	I/II	NCT02547961
	Sarcoma	I	NCT00902044
IL-13Rα2	Glioma, Glioblastoma	I	NCT02208362
Mesothelin	Pancreatic	I/II	NCT02959151
	Cervical	I/II	NCT03356795
	Advanced Solid Tumors	I/II	NCT03615313
	Pancreatic, Ovarian, Mesothelioma	I	NCT02159716
	Malignant Pleural Disease, Mesothelioma, Lung, Breast	I	NCT02414269
MUC1	Cervical	I/II	NCT03356795
	Esophageal	I/II	NCT03706326
	Non-Small Cell Lung	I/II	NCT03525782
	Osteoid Sarcoma, Ewing Sarcoma	I/II	NCT03356782
	Intrahepatic Cholangiocarcinoma	I/II	NCT03633773
MUC-16	Ovarian	I	NCT02498912
PSMA	Urothelial Bladder	I/II	NCT03185468
	Cervical	I/II	NCT03356795
	Prostate	I	NCT01140373

6. Conclusions

The late 19th century observation that tumors could be treated with cocktails of heat-killed bacteria has proven highly influential. Unbeknownst to William Coley and his contemporaries, this would prove to be one of the first documented cases of tumor regression by induced activation of the immune system. Coley's legacy would help spur subsequent hypotheses of immunosurveillance mechanisms capable of mediating steady-state tumor recognition and elimination. Over the next century, exploitation of these mechanisms was to become a major priority as immunotherapies continued to evolve.

Treatment regimens using recombinant cytokines that activate immune cell proliferation and effector functions are efficacious in treating selected patient populations. Likewise, strategies employing immune checkpoint blockade against tumor cells that express co-inhibitory molecules have reached clinical milestones once thought to be unachievable. Vaccines against tumor-associated antigens have demonstrated clinical benefit, with applications turning towards neoantigens as patient-specific tumor sequencing becomes feasible. FDA approval of the oncolytic virotherapy, T-VEC, may also offer another option for locally-acting immune stimulation and antitumor activity when resection, chemotherapy, or radiation are not amenable. Moreover, the success of CAR-T cell therapy in patients with hematological malignancies has established adoptive cell therapy as a viable treatment modality. As the costs associated with patient and tumor genome sequencing continue to decrease, the rapidly evolving "omics"-level of data acquisition and processing may enable precise treatment strategies for these patients. Deconvoluting patient-specific tumor heterogeneity with the assistance of "big data" may enable clinicians and researchers to select the best candidate immunotherapy from the start, while taking proactive approaches to overcome resistance mechanisms in an adaptive tumor microenvironment [198]. The previous ~150 years of immuno-oncology research without significant clinical success has now enabled "hockey stick" growth in exploration of the safety and efficacy of immune-centric therapies in clinical trials. Just as William Coley's fundamental discoveries have shaped modern cancer immunotherapy, so too shall current efforts influence the future of cancer treatment.

Author Contributions: R.D.C., J.C.F.J., and A.E.S. conceived the review, R.D.C. and J.C.F.J. wrote the manuscript, and A.E.S. revised. All authors have read and agreed to the published version of the manuscript.

Funding: The authors are supported, in part, by the Department of Defense Congressionally Directed Medical Research Programs (#W81XWH-17-1-0299, #W81XWH-19-1-0263, and #W81XWH-19-1-0067 to A.E.S.), Targeted Diagnostics and Therapeutics Inc. (A.E.S.), the Degregorio Family Foundation (A.E.S.), and a Pre-Doctoral Fellowship in Pharmacology/Toxicology from the PhRMA Foundation (J.C.F.). Figure 1 was created with BioRender.com.

Conflicts of Interest: The authors declare no conflict of interest.

References

1. Couzin-Frankel, J. Breakthrough of the year 2013. Cancer immunotherapy. *Science* **2013**, *342*, 1432–1433. [CrossRef] [PubMed]
2. Challis, G.B.; Stam, H.J. The spontaneous regression of cancer. A review of cases from 1900 to 1987. *Acta Oncol.* **1990**, *29*, 545–550. [CrossRef] [PubMed]
3. Papac, R.J. Spontaneous regression of cancer. *Cancer Treat. Rev.* **1996**, *22*, 395–423. [CrossRef]
4. Smith, J.L.; Stehlin, J.S. Spontaneous regression of primary malignant melanomas with regional metastases. *Cancer* **1965**, *18*, 1399–1415. [CrossRef]
5. Haridy, Y.; Witzmann, F.; Asbach, P.; Schoch, R.R.; Fröbisch, N.; Rothschild, B.M. Triassic Cancer-Osteosarcoma in a 240-Million-Year-Old Stem-Turtle. *JAMA Oncol.* **2019**, *5*, 425–426. [CrossRef] [PubMed]
6. Coley, W.B., II. contribution to the knowledge of sarcoma. *Ann. Surg.* **1891**, *14*, 199–220. [CrossRef] [PubMed]
7. Coley, W.B. The treatment of malignant tumors by repeated inoculations of erysipelas: With a report of ten original cases. 1. *Am. J. Med. Sci.* **1893**, *105*, 487. [CrossRef]
8. Stevens, D.L.; Bryant, A.E. Impetigo, erysipelas and cellulitis. In *Streptococcus Pyogenes: Basic Biology to Clinical Manifestations*; Ferretti, J.J., Stevens, D.L., Fischetti, V.A., Eds.; University of Oklahoma Health Sciences Center: Oklahoma City, OK, USA, 2016.

9.   Fehleisen, F. *Die Aetiologie des Erysipels*; Theodor Fischer: Berlin, Germany, 1883.

10.   Cann, S.A.H.; Van Netten, J.P.; Van Netten, C. Dr William Coley and tumour regression: A place in history or in the future. *Postgrad. Med.* **2003**, *79*, 672–680.

11.   McCarthy, E.F. The toxins of William B. Coley and the treatment of bone and soft-tissue sarcomas. *Iowa Orthop. J.* **2006**, *26*, 154–158.

12.   Coley, W.B. The therapeutic value of the mixed toxins of the streptococcus of erysipelas and bacillus prodigiosus in the treatment of one hundred and sixty cases. *Am. J. Med. Sci.* **1896**, *112*, 251. [CrossRef]

13.   Coley, W.B. The Treatment of Inoperable Sarcoma by Bacterial Toxins (the Mixed Toxins of the Streptococcus erysipelas and the Bacillus prodigiosus). *Proc. R. Soc. Med.* **1910**, *3*, 1–48. [CrossRef] [PubMed]

14.   Johnston, B.J.; Novales, E.T. Clinical effect of Coley's toxin. II. A seven-year study. *Cancer Chemother. Rep.* **1962**, *21*, 43–68. [PubMed]

15.   Ehrlich, P. Über den jetzigen Stand der Karzinomforschung. *Ned Tijdschr Geneeskd.* **1909**, *5*, 273–290.

16.   Burnet, F.M. The concept of immunological surveillance. *Immunol. Asp. Neoplasia* **1970**, *13*, 1–27.

17.   Thomas, L. On immunosurveillance in human cancer. *Yale J. Biol. Med.* **1982**, *55*, 329–333.

18.   Gross, L. Intradermal immunization of C3H mice against a sarcoma that originated in an animal of the same line. *Cancer Res.* **1943**, *3*, 326–333.

19.   Prehn, R.T.; Main, J.M. Immunity to methylcholanthrene-induced sarcomas. *J. Natl. Cancer Inst.* **1957**, *18*, 769–778.

20.   Shinkai, Y.; Rathbun, G.; Lam, K.P.; Oltz, E.M.; Stewart, V.; Mendelsohn, M.; Charron, J.; Datta, M.; Young, F.; Stall, A.M. RAG-2-deficient mice lack mature lymphocytes owing to inability to initiate V(D)J rearrangement. *Cell* **1992**, *68*, 855–867. [CrossRef]

21.   Shankaran, V.; Ikeda, H.; Bruce, A.T.; White, J.M.; Swanson, P.E.; Old, L.J.; Schreiber, R.D. IFNgamma and lymphocytes prevent primary tumour development and shape tumour immunogenicity. *Nature* **2001**, *410*, 1107–1111. [CrossRef]

22.   Cretney, E.; Takeda, K.; Yagita, H.; Glaccum, M.; Peschon, J.J.; Smyth, M.J. Increased susceptibility to tumor initiation and metastasis in TNF-related apoptosis-inducing ligand-deficient mice. *J. Immunol.* **2002**, *168*, 1356–1361. [CrossRef]

23.   van den Broek, M.E.; Kägi, D.; Ossendorp, F.; Toes, R.; Vamvakas, S.; Lutz, W.K.; Melief, C.J.; Zinkernagel, R.M.; Hengartner, H. Decreased tumor surveillance in perforin-deficient mice. *J. Exp. Med.* **1996**, *184*, 1781–1790. [CrossRef] [PubMed]

24.   Mayor, P.C.; Eng, K.H.; Singel, K.L.; Abrams, S.I.; Odunsi, K.; Moysich, K.B.; Fuleihan, R.; Garabedian, E.; Lugar, P.; Ochs, H.D.; et al. Cancer in primary immunodeficiency diseases: Cancer incidence in the United States Immune Deficiency Network Registry. *J. Allergy Clin. Immunol.* **2018**, *141*, 1028–1035. [CrossRef] [PubMed]

25.   Engels, E.A.; Pfeiffer, R.M.; Fraumeni, J.F.; Kasiske, B.L.; Israni, A.K.; Snyder, J.J.; Wolfe, R.A.; Goodrich, N.P.; Bayakly, A.R.; Clarke, C.A.; et al. Spectrum of cancer risk among US solid organ transplant recipients. *JAMA* **2011**, *306*, 1891–1901. [CrossRef] [PubMed]

26.   Goedert, J.J.; Coté, T.R.; Virgo, P.; Scoppa, S.M.; Kingma, D.W.; Gail, M.H.; Jaffe, E.S.; Biggar, R.J. Spectrum of AIDS-associated malignant disorders. *Lancet* **1998**, *351*, 1833–1839. [CrossRef]

27.   Schreiber, R.D.; Old, L.J.; Smyth, M.J. Cancer immunoediting: Integrating immunity's roles in cancer suppression and promotion. *Science* **2011**, *331*, 1565–1570. [CrossRef] [PubMed]

28.   Koebel, C.M.; Vermi, W.; Swann, J.B.; Zerafa, N.; Rodig, S.J.; Old, L.J.; Smyth, M.J.; Schreiber, R.D. Adaptive immunity maintains occult cancer in an equilibrium state. *Nature* **2007**, *450*, 903–907. [CrossRef]

29.   DuPage, M.; Mazumdar, C.; Schmidt, L.M.; Cheung, A.F.; Jacks, T. Expression of tumour-specific antigens underlies cancer immunoediting. *Nature* **2012**, *482*, 405–409. [CrossRef]

30.   Miescher, S.; Whiteside, T.L.; Moretta, L.; von Fliedner, V. Clonal and frequency analyses of tumor-infiltrating T lymphocytes from human solid tumors. *J. Immunol.* **1987**, *138*, 4004–4011.

31.   Li, Y.; Patel, S.P.; Roszik, J.; Qin, Y. Hypoxia-Driven Immunosuppressive Metabolites in the Tumor Microenvironment: New Approaches for Combinational Immunotherapy. *Front. Immunol.* **2018**, *9*, 1591. [CrossRef]

32.   Barsoum, I.B.; Koti, M.; Siemens, D.R.; Graham, C.H. Mechanisms of hypoxia-mediated immune escape in cancer. *Cancer Res.* **2014**, *74*, 7185–7190. [CrossRef]

33. Fischer, K.; Hoffmann, P.; Voelkl, S.; Meidenbauer, N.; Ammer, J.; Edinger, M.; Gottfried, E.; Schwarz, S.; Rothe, G.; Hoves, S.; et al. Inhibitory effect of tumor cell-derived lactic acid on human T cells. *Blood* **2007**, *109*, 3812–3819. [CrossRef] [PubMed]

34. Togashi, Y.; Shitara, K.; Nishikawa, H. Regulatory T cells in cancer immunosuppression—Implications for anticancer therapy. *Nat. Rev. Clin. Oncol.* **2019**, *16*, 356–371. [CrossRef] [PubMed]

35. Butt, A.Q.; Mills, K.H.G. Immunosuppressive networks and checkpoints controlling antitumor immunity and their blockade in the development of cancer immunotherapeutics and vaccines. *Oncogene* **2014**, *33*, 4623–4631. [CrossRef] [PubMed]

36. Sakaguchi, S.; Yamaguchi, T.; Nomura, T.; Ono, M. Regulatory T cells and immune tolerance. *Cell* **2008**, *133*, 775–787. [CrossRef]

37. Gery, I.; Gershon, R.K.; Waksman, B.H. Potentiation of the T-lymphocyte response to mitogens. I. The responding cell. *J. Exp. Med.* **1972**, *136*, 128–142. [CrossRef]

38. Mier, J.W.; Gallo, R.C. Purification and some characteristics of human T-cell growth factor from phytohemagglutinin-stimulated lymphocyte-conditioned media. *Proc. Natl. Acad. Sci. USA* **1980**, *77*, 6134–6138. [CrossRef]

39. Welte, K.; Wang, C.Y.; Mertelsmann, R.; Venuta, S.; Feldman, S.P.; Moore, M.A. Purification of human interleukin 2 to apparent homogeneity and its molecular heterogeneity. *J. Exp. Med.* **1982**, *156*, 454–464. [CrossRef]

40. Rosenberg, S.A.; Mulé, J.J.; Spiess, P.J.; Reichert, C.M.; Schwarz, S.L. Regression of established pulmonary metastases and subcutaneous tumor mediated by the systemic administration of high-dose recombinant interleukin 2. *J. Exp. Med.* **1985**, *161*, 1169–1188. [CrossRef]

41. Eberlein, T.J.; Rosenstein, M.; Rosenberg, S.A. Regression of a disseminated syngeneic solid tumor by systemic transfer of lymphoid cells expanded in interleukin 2. *J. Exp. Med.* **1982**, *156*, 385–397. [CrossRef]

42. Rosenberg, S.A.; Lotze, M.T.; Muul, L.M.; Leitman, S.; Chang, A.E.; Ettinghausen, S.E.; Matory, Y.L.; Skibber, J.M.; Shiloni, E.; Vetto, J.T. Observations on the systemic administration of autologous lymphokine-activated killer cells and recombinant interleukin-2 to patients with metastatic cancer. *N. Engl. J. Med.* **1985**, *313*, 1485–1492. [CrossRef]

43. Rosenberg, S.A.; Lotze, M.T.; Muul, L.M.; Chang, A.E.; Avis, F.P.; Leitman, S.; Linehan, W.M.; Robertson, C.N.; Lee, R.E.; Rubin, J.T. A progress report on the treatment of 157 patients with advanced cancer using lymphokine-activated killer cells and interleukin-2 or high-dose interleukin-2 alone. *N. Engl. J. Med.* **1987**, *316*, 889–897. [CrossRef] [PubMed]

44. Schwartz, R.N.; Stover, L.; Dutcher, J. Managing toxicities of high-dose interleukin-2. *Oncology* **2002**, *16*, 11–20. [PubMed]

45. Rosenberg, S.A.; Yang, J.C.; Topalian, S.L.; Schwartzentruber, D.J.; Weber, J.S.; Parkinson, D.R.; Seipp, C.A.; Einhorn, J.H.; White, D.E. Treatment of 283 consecutive patients with metastatic melanoma or renal cell cancer using high-dose bolus interleukin 2. *JAMA* **1994**, *271*, 907–913. [CrossRef] [PubMed]

46. Dillman, R.O.; Church, C.; Oldham, R.K.; West, W.H.; Schwartzberg, L.; Birch, R. Inpatient continuous-infusion interleukin-2 in 788 patients with cancer. The National Biotherapy Study Group experience. *Cancer* **1993**, *71*, 2358–2370. [CrossRef]

47. Lee, S.; Margolin, K. Cytokines in cancer immunotherapy. *Cancers* **2011**, *3*, 3856–3893. [CrossRef]

48. Ku, C.C.; Murakami, M.; Sakamoto, A.; Kappler, J.; Marrack, P. Control of homeostasis of CD8+ memory T cells by opposing cytokines. *Science* **2000**, *288*, 675–678. [CrossRef]

49. Evans, R.; Fuller, J.A.; Christianson, G.; Krupke, D.M. IL-15 mediates anti-tumor effects after cyclophosphamide injection of tumor-bearing mice and enhances adoptive immunotherapy: The potential role of NK cell. *Cell. Immunol.* **1997**, *178*, 197. [CrossRef]

50. Kobayashi, H.; Dubois, S.; Sato, N.; Sabzevari, H.; Sakai, Y.; Waldmann, T.A.; Tagaya, Y. Role of trans-cellular IL-15 presentation in the activation of NK cell-mediated killing, which leads to enhanced tumor immunosurveillance. *Blood* **2005**, *105*, 721–727. [CrossRef]

51. Waldmann, T.A. The biology of interleukin-2 and interleukin-15: Implications for cancer therapy and vaccine design. *Nat. Rev. Immunol.* **2006**, *6*, 595–601. [CrossRef]

52. Bessard, A.; Solé, V.; Bouchaud, G.; Quéméner, A. High antitumor activity of RLI, an interleukin-15 (IL-15)–IL-15 receptor α fusion protein, in metastatic melanoma and colorectal cancer. *Mol. Cancer Ther.* **2009**, *8*, 2736–2745. [CrossRef]

53. Howie, J.W. Experiments with interferon in man: A report to the medical research council from the scientific committee on interferon. *Lancet* **1965**, *1*, 505. [PubMed]

54. Müller, U.; Steinhoff, U.; Reis, L.F.; Hemmi, S.; Pavlovic, J.; Zinkernagel, R.M.; Aguet, M. Functional role of type I and type II interferons in antiviral defense. *Science* **1994**, *264*, 1918–1921. [CrossRef]

55. Waldmann, T.A. Cytokines in cancer immunotherapy. *Cold Spring Harb. Perspect. Biol.* **2017**, *10*. [CrossRef] [PubMed]

56. Cantell, K.; Hirvonen, S.; Koistinen, V. [71] Partial purification of human leukocyte interferon on a large scale. In *Methods in Enzymology*; Academic Press: Cambridge, MA, USA, 1981.

57. Merigan, T.C.; Sikora, K.; Breeden, J.H.; Levy, R.; Rosenberg, S.A. Preliminary observations on the effect of human leukocyte interferon in non-Hodgkin's lymphoma. *N. Engl. J. Med.* **1978**, *299*, 1449–1453. [CrossRef] [PubMed]

58. Quesada, J.R.; Hersh, E.M.; Gutterman, J.U. Hairy cell leukemia: Induction of remission with alpha interferon. *Blood* **1983**, *62* (Suppl. 1), 207. [CrossRef] [PubMed]

59. Quesada, J.R.; Swanson, D.A.; Trindade, A.; Gutterman, J.U. Renal cell carcinoma: Antitumor effects of leukocyte interferon. *Cancer Res.* **1983**, *43*, 940–947.

60. Krown, S.E.; Burk, M.W.; Kirkwood, J.M.; Kerr, D. Human leukocyte (alpha) interferon in metastatic malignant melanoma. *Am. Cancer Soc. Phase II Trial* **1984**, *68*, 723–726.

61. Kirkwood, J.M.; Ibrahim, J.G.; Sondak, V.K.; Richards, J.; Flaherty, L.E.; Ernstoff, M.S.; Smith, T.J.; Rao, U.; Steele, M.; Blum, R.H. High-and low-dose interferon alfa-2b in high-risk melanoma: First analysis of intergroup trial E1690/S9111/C9190. *J. Clin. Oncol.* **2000**, *18*, 2444–2458. [CrossRef]

62. Hercus, T.R.; Thomas, D.; Guthridge, M.A.; Ekert, P.G.; King-Scott, J.; Parker, M.W.; Lopez, A.F. The granulocyte-macrophage colony-stimulating factor receptor: Linking its structure to cell signaling and its role in disease. *Blood* **2009**, *114*, 1289–1298. [CrossRef]

63. Dranoff, G.; Jaffee, E.; Lazenby, A.; Golumbek, P.; Levitsky, H.; Brose, K.; Jackson, V.; Hamada, H.; Pardoll, D.; Mulligan, R.C. Vaccination with irradiated tumor cells engineered to secrete murine granulocyte-macrophage colony-stimulating factor stimulates potent, specific, and long-lasting anti-tumor immunity. *Proc. Natl. Acad. Sci. USA* **1993**, *90*, 3539–3543. [CrossRef]

64. Soiffer, R.; Lynch, T.; Mihm, M.; Jung, K.; Rhuda, C.; Schmollinger, J.C.; Hodi, F.S.; Liebster, L.; Lam, P.; Mentzer, S.; et al. Vaccination with irradiated autologous melanoma cells engineered to secrete human granulocyte-macrophage colony-stimulating factor generates potent antitumor immunity in patients with metastatic melanoma. *Proc. Natl. Acad. Sci. USA* **1998**, *95*, 13141–13146. [CrossRef] [PubMed]

65. Small, E.J.; Reese, D.M.; Um, B.; Whisenant, S.; Dixon, S.C.; Figg, W.D. Therapy of advanced prostate cancer with granulocyte macrophage colony-stimulating factor. *Clin. Cancer Res.* **1999**, *5*, 1738–1744. [PubMed]

66. Lawson, D.H.; Lee, S.; Zhao, F.; Tarhini, A.A.; Margolin, K.A.; Ernstoff, M.S.; Atkins, M.B.; Cohen, G.I.; Whiteside, T.L.; Butterfield, L.H.; et al. Randomized, Placebo-Controlled, Phase III Trial of Yeast-Derived Granulocyte-Macrophage Colony-Stimulating Factor (GM-CSF) Versus Peptide Vaccination Versus GM-CSF Plus Peptide Vaccination Versus Placebo in Patients With No Evidence of Disease After Complete Surgical Resection of Locally Advanced and/or Stage IV Melanoma: A Trial of the Eastern Cooperative Oncology Group-American College of Radiology Imaging Network Cancer Research Group (E4697). *J. Clin. Oncol.* **2015**, *33*, 4066–4076. [CrossRef] [PubMed]

67. Hodi, F.S.; Lee, S.; McDermott, D.F.; Rao, U.N.; Butterfield, L.H.; Tarhini, A.A.; Leming, P.; Puzanov, I.; Shin, D.; Kirkwood, J.M. Ipilimumab plus sargramostim vs ipilimumab alone for treatment of metastatic melanoma: A randomized clinical trial. *JAMA* **2014**, *312*, 1744–1753. [CrossRef]

68. O'Day, S.J.; Atkins, M.B.; Boasberg, P. Phase II multicenter trial of maintenance biotherapy after induction concurrent Biochemotherapy for patients with metastatic melanoma. *J. Clin. Oncol.* **2009**, *27*, 6207–6212. [CrossRef]

69. Tang, H.; Qiao, J.; Fu, Y.-X. Immunotherapy and tumor microenvironment. *Cancer Lett.* **2016**, *370*, 85–90. [CrossRef]

70. Mortarini, R.; Piris, A.; Maurichi, A.; Molla, A.; Bersani, I.; Bono, A.; Bartoli, C.; Santinami, M.; Lombardo, C.; Ravagnani, F.; et al. Lack of terminally differentiated tumor-specific CD8+ T cells at tumor site in spite of antitumor immunity to self-antigens in human metastatic melanoma. *Cancer Res.* **2003**, *63*, 2535–2545.

71. Deng, L.; Zhang, H.; Luan, Y.; Zhang, J.; Xing, Q.; Dong, S.; Wu, X.; Liu, M.; Wang, S. Accumulation of foxp3+ T regulatory cells in draining lymph nodes correlates with disease progression and immune suppression in colorectal cancer patients. *Clin. Cancer Res.* **2010**, *16*, 4105–4112. [CrossRef]

72. Gallimore, A.; Glithero, A.; Godkin, A.; Tissot, A.C.; Plückthun, A.; Elliott, T.; Hengartner, H.; Zinkernagel, R. Induction and exhaustion of lymphocytic choriomeningitis virus-specific cytotoxic T lymphocytes visualized using soluble tetrameric major histocompatibility complex class I-peptide complexes. *J. Exp. Med.* **1998**, *187*, 1383–1393. [CrossRef]

73. Joffre, O.; Nolte, M.A.; Spörri, R.; Reis e Sousa, C. Inflammatory signals in dendritic cell activation and the induction of adaptive immunity. *Immunol. Rev.* **2009**, *227*, 234–247. [CrossRef]

74. Shahinian, A.; Pfeffer, K.; Lee, K.P.; Kündig, T.M.; Kishihara, K.; Wakeham, A.; Kawai, K.; Ohashi, P.S.; Thompson, C.B.; Mak, T.W. Differential T cell costimulatory requirements in CD28-deficient mice. *Science* **1993**, *261*, 609–612. [CrossRef] [PubMed]

75. Chen, L.; Flies, D.B. Molecular mechanisms of T cell co-stimulation and co-inhibition. *Nat. Rev. Immunol.* **2013**, *13*, 227–242. [CrossRef] [PubMed]

76. Hargadon, K.M.; Johnson, C.E.; Williams, C.J. Immune checkpoint blockade therapy for cancer: An overview of FDA-approved immune checkpoint inhibitors. *Int. Immunopharmacol.* **2018**, *62*, 29–39. [CrossRef] [PubMed]

77. Brunet, J.F.; Denizot, F.; Luciani, M.F.; Roux-Dosseto, M.; Suzan, M.; Mattei, M.G.; Golstein, P. A new member of the immunoglobulin superfamily—CTLA-4. *Nature* **1987**, *328*, 267–270. [CrossRef]

78. Walunas, T.L.; Lenschow, D.J.; Bakker, C.Y.; Linsley, P.S.; Freeman, G.J.; Green, J.M.; Thompson, C.B.; Bluestone, J.A. CTLA-4 can function as a negative regulator of T cell activation. *Immunity* **1994**, *1*, 405–413. [CrossRef]

79. Qureshi, O.S.; Zheng, Y.; Nakamura, K.; Attridge, K.; Manzotti, C.; Schmidt, E.M.; Baker, J.; Jeffery, L.E.; Kaur, S.; Briggs, Z.; et al. Trans-endocytosis of CD80 and CD86: A molecular basis for the cell-extrinsic function of CTLA-4. *Science* **2011**, *332*, 600–603. [CrossRef]

80. Krummel, M.F.; Allison, J.P. CD28 and CTLA-4 have opposing effects on the response of T cells to stimulation. *J. Exp. Med.* **1995**, *182*, 459–465. [CrossRef]

81. Tivol, E.A.; Borriello, F.; Schweitzer, A.N.; Lynch, W.P.; Bluestone, J.A.; Sharpe, A.H. Loss of CTLA-4 leads to massive lymphoproliferation and fatal multiorgan tissue destruction, revealing a critical negative regulatory role of CTLA-4. *Immunity* **1995**, *3*, 541–547. [CrossRef]

82. Waterhouse, P.; Penninger, J.M.; Timms, E.; Wakeham, A.; Shahinian, A.; Lee, K.P.; Thompson, C.B.; Griesser, H.; Mak, T.W. Lymphoproliferative disorders with early lethality in mice deficient in Ctla-4. *Science* **1995**, *270*, 985–988. [CrossRef]

83. Leach, D.R.; Krummel, M.F.; Allison, J.P. Enhancement of antitumor immunity by CTLA-4 blockade. *Science* **1996**, *271*, 1734–1736. [CrossRef]

84. Hodi, F.S.; O'Day, S.J.; McDermott, D.F.; Weber, R.W.; Sosman, J.A.; Haanen, J.B.; Gonzalez, R.; Robert, C.; Schadendorf, D.; Hassel, J.C.; et al. Improved survival with ipilimumab in patients with metastatic melanoma. *N. Engl. J. Med.* **2010**, *363*, 711–723. [CrossRef] [PubMed]

85. Robert, C.; Thomas, L.; Bondarenko, I.; O'Day, S.; Weber, J.; Garbe, C.; Lebbe, C.; Baurain, J.-F.; Testori, A.; Grob, J.-J.; et al. Ipilimumab plus dacarbazine for previously untreated metastatic melanoma. *N. Engl. J. Med.* **2011**, *364*, 2517–2526. [CrossRef] [PubMed]

86. Schadendorf, D.; Hodi, F.S.; Robert, C.; Weber, J.S.; Margolin, K.; Hamid, O.; Patt, D.; Chen, T.-T.; Berman, D.M.; Wolchok, J.D. Pooled Analysis of Long-Term Survival Data From Phase II and Phase III Trials of Ipilimumab in Unresectable or Metastatic Melanoma. *J. Clin. Oncol.* **2015**, *33*, 1889–1894. [CrossRef] [PubMed]

87. Darvin, P.; Toor, S.M.; Sasidharan Nair, V.; Elkord, E. Immune checkpoint inhibitors: Recent progress and potential biomarkers. *Exp. Mol. Med.* **2018**, *50*, 165. [CrossRef] [PubMed]

88. Ishida, Y.; Agata, Y.; Shibahara, K.; Honjo, T. Induced expression of PD-1, a novel member of the immunoglobulin gene superfamily, upon programmed cell death. *EMBO J.* **1992**, *11*, 3887–3895. [CrossRef] [PubMed]

89. Nishimura, H.; Nose, M.; Hiai, H.; Minato, N.; Honjo, T. Development of lupus-like autoimmune diseases by disruption of the PD-1 gene encoding an ITIM motif-carrying immunoreceptor. *Immunity* **1999**, *11*, 141–151. [CrossRef]

90. Dong, H.; Zhu, G.; Tamada, K.; Chen, L. B7-H1, a third member of the B7 family, co-stimulates T-cell proliferation and interleukin-10 secretion. *Nat. Med.* **1999**, *5*, 1365–1369. [CrossRef]

91. Freeman, G.J.; Long, A.J.; Iwai, Y.; Bourque, K.; Chernova, T.; Nishimura, H.; Fitz, L.J.; Malenkovich, N.; Okazaki, T.; Byrne, M.C.; et al. Engagement of the PD-1 immunoinhibitory receptor by a novel B7 family member leads to negative regulation of lymphocyte activation. *J. Exp. Med.* **2000**, *192*, 1027–1034. [CrossRef]

92. Iwai, Y.; Ishida, M.; Tanaka, Y.; Okazaki, T.; Honjo, T.; Minato, N. Involvement of PD-L1 on tumor cells in the escape from host immune system and tumor immunotherapy by PD-L1 blockade. *Proc. Natl. Acad. Sci. USA* **2002**, *99*, 12293–12297. [CrossRef]

93. Curiel, T.J.; Wei, S.; Dong, H.; Alvarez, X.; Cheng, P.; Mottram, P.; Krzysiek, R.; Knutson, K.L.; Daniel, B.; Zimmermann, M.C.; et al. Blockade of B7-H1 improves myeloid dendritic cell-mediated antitumor immunity. *Nat. Med.* **2003**, *9*, 562–567. [CrossRef]

94. Strome, S.E.; Dong, H.; Tamura, H.; Voss, S.G.; Flies, D.B.; Tamada, K.; Salomao, D.; Cheville, J.; Hirano, F.; Lin, W.; et al. B7-H1 blockade augments adoptive T-cell immunotherapy for squamous cell carcinoma. *Cancer Res.* **2003**, *63*, 6501–6505.

95. Iwai, Y.; Terawaki, S.; Honjo, T. PD-1 blockade inhibits hematogenous spread of poorly immunogenic tumor cells by enhanced recruitment of effector T cells. *Int. Immunol.* **2005**, *17*, 133–144. [CrossRef] [PubMed]

96. Hamanishi, J.; Mandai, M.; Iwasaki, M.; Okazaki, T.; Tanaka, Y.; Yamaguchi, K.; Higuchi, T.; Yagi, H.; Takakura, K.; Minato, N.; et al. Programmed cell death 1 ligand 1 and tumor-infiltrating CD8+ T lymphocytes are prognostic factors of human ovarian cancer. *Proc. Natl. Acad. Sci. USA* **2007**, *104*, 3360–3365. [CrossRef] [PubMed]

97. Brahmer, J.R.; Tykodi, S.S.; Chow, L.Q.M.; Hwu, W.-J.; Topalian, S.L.; Hwu, P.; Drake, C.G.; Camacho, L.H.; Kauh, J.; Odunsi, K.; et al. Safety and activity of anti-PD-L1 antibody in patients with advanced cancer. *N. Engl. J. Med.* **2012**, *366*, 2455–2465. [CrossRef] [PubMed]

98. Topalian, S.L.; Hodi, F.S.; Brahmer, J.R.; Gettinger, S.N.; Smith, D.C.; McDermott, D.F.; Powderly, J.D.; Carvajal, R.D.; Sosman, J.A.; Atkins, M.B.; et al. Safety, activity, and immune correlates of anti-PD-1 antibody in cancer. *N. Engl. J. Med.* **2012**, *366*, 2443–2454. [CrossRef]

99. Garon, E.B.; Rizvi, N.A.; Hui, R.; Leighl, N.; Balmanoukian, A.S.; Eder, J.P.; Patnaik, A.; Aggarwal, C.; Gubens, M.; Horn, L.; et al. KEYNOTE-001 Investigators Pembrolizumab for the treatment of non-small-cell lung cancer. *N. Engl. J. Med.* **2015**, *372*, 2018–2028. [CrossRef]

100. Weber, J.S.; D'Angelo, S.P.; Minor, D.; Hodi, F.S.; Gutzmer, R.; Neyns, B.; Hoeller, C.; Khushalani, N.I.; Miller, W.H.; Lao, C.D.; et al. Nivolumab versus chemotherapy in patients with advanced melanoma who progressed after anti-CTLA-4 treatment (CheckMate 037): A randomised, controlled, open-label, phase 3 trial. *Lancet Oncol.* **2015**, *16*, 375–384. [CrossRef]

101. Schachter, J.; Ribas, A.; Long, G.V.; Arance, A.; Grob, J.-J.; Mortier, L.; Daud, A.; Carlino, M.S.; McNeil, C.; Lotem, M.; et al. Pembrolizumab versus ipilimumab for advanced melanoma: Final overall survival results of a multicentre, randomised, open-label phase 3 study (KEYNOTE-006). *Lancet* **2017**, *390*, 1853–1862. [CrossRef]

102. Powles, T.; Durán, I.; van der Heijden, M.S.; Loriot, Y.; Vogelzang, N.J.; De Giorgi, U.; Oudard, S.; Retz, M.M.; Castellano, D.; Bamias, A.; et al. Atezolizumab versus chemotherapy in patients with platinum-treated locally advanced or metastatic urothelial carcinoma (IMvigor211): A multicentre, open-label, phase 3 randomised controlled trial. *Lancet* **2018**, *391*, 748–757. [CrossRef]

103. Rittmeyer, A.; Barlesi, F.; Waterkamp, D.; Park, K.; Ciardiello, F.; von Pawel, J.; Gadgeel, S.M.; Hida, T.; Kowalski, D.M.; Dols, M.C.; et al. OAK Study Group Atezolizumab versus docetaxel in patients with previously treated non-small-cell lung cancer (OAK): A phase 3, open-label, multicentre randomised controlled trial. *Lancet* **2017**, *389*, 255–265. [CrossRef]

104. Narayan, P.; Wahby, S.; Gao, J.J.; Amiri-Kordestani, L.; Ibrahim, A.; Bloomquist, E.; Tang, S.; Xu, Y.; Liu, J.; Fu, W.; et al. FDA Approval Summary: Atezolizumab plus paclitaxel protein-bound for the treatment of patients with advanced or metastatic TNBC whose tumors express PD-L1. *Clin. Cancer Res.* **2020**. [CrossRef] [PubMed]

105. Kaufman, H.L.; Russell, J.; Hamid, O.; Bhatia, S.; Terheyden, P.; D'Angelo, S.P.; Shih, K.C.; Lebbé, C.; Linette, G.P.; Milella, M.; et al. Avelumab in patients with chemotherapy-refractory metastatic Merkel cell carcinoma: A multicentre, single-group, open-label, phase 2 trial. *Lancet Oncol.* **2016**, *17*, 1374–1385. [CrossRef]

106. Motzer, R.J.; Tannir, N.M.; McDermott, D.F.; Arén Frontera, O.; Melichar, B.; Choueiri, T.K.; Plimack, E.R.; Barthélémy, P.; Porta, C.; George, S.; et al. CheckMate 214 Investigators Nivolumab plus Ipilimumab versus Sunitinib in Advanced Renal-Cell Carcinoma. *N. Engl. J. Med.* **2018**, *378*, 1277–1290. [CrossRef]

107. Jenkins, R.W.; Barbie, D.A.; Flaherty, K.T. Mechanisms of resistance to immune checkpoint inhibitors. *Br. J. Cancer* **2018**, *118*, 9–16. [CrossRef] [PubMed]

108. Anderson, A.C.; Joller, N.; Kuchroo, V.K. Lag-3, Tim-3, and TIGIT: Co-inhibitory Receptors with Specialized Functions in Immune Regulation. *Immunity* **2016**, *44*, 989–1004. [CrossRef] [PubMed]

109. Routy, B.; Le Chatelier, E.; Derosa, L.; Duong, C.P.M.; Alou, M.T.; Daillère, R.; Fluckiger, A.; Messaoudene, M.; Rauber, C.; Roberti, M.P.; et al. Gut microbiome influences efficacy of PD-1-based immunotherapy against epithelial tumors. *Science* **2018**, *359*, 91–97. [CrossRef]

110. Gopalakrishnan, V.; Spencer, C.N.; Nezi, L.; Reuben, A.; Andrews, M.C.; Karpinets, T.V.; Prieto, P.A.; Vicente, D.; Hoffman, K.; Wei, S.C.; et al. Gut microbiome modulates response to anti-PD-1 immunotherapy in melanoma patients. *Science* **2018**, *359*, 97–103. [CrossRef]

111. Hanna, M.G.; Peters, L.C. Specific immunotherapy of established visceral micrometastases by BCG-tumor cell vaccine alone or as an adjunct to surgery. *Cancer* **1978**, *42*, 2613–2625. [CrossRef]

112. Murray, D.R.; Cassel, W.A.; Torbin, A.H.; Olkowski, Z.L.; Moore, M.E. Viral oncolysate in the management of malignant melanoma. II. Clinical studies. *Cancer* **1977**, *40*, 680–686. [CrossRef]

113. McIllmurray, M.B.; Embleton, M.J.; Reeves, W.G.; Langman, M.J.; Deane, M. Controlled trial of active immunotherapy in management of stage IIB malignant melanoma. *Br. Med. J.* **1977**, *1*, 540–542. [CrossRef]

114. van der Bruggen, P.; Traversari, C.; Chomez, P.; Lurquin, C.; De Plaen, E.; Van den Eynde, B.; Knuth, A.; Boon, T. A gene encoding an antigen recognized by cytolytic T lymphocytes on a human melanoma. *Science* **1991**, *254*, 1643–1647. [CrossRef] [PubMed]

115. Cheever, M.A.; Allison, J.P.; Ferris, A.S.; Finn, O.J.; Hastings, B.M.; Hecht, T.T.; Mellman, I.; Prindiville, S.A.; Viner, J.L.; Weiner, L.M.; et al. The prioritization of cancer antigens: A national cancer institute pilot project for the acceleration of translational research. *Clin. Cancer Res.* **2009**, *15*, 5323–5337. [CrossRef] [PubMed]

116. Thomas, R.; Al-Khadairi, G.; Roelands, J.; Hendrickx, W.; Dermime, S.; Bedognetti, D.; Decock, J. NY-ESO-1 Based Immunotherapy of Cancer: Current Perspectives. *Front. Immunol.* **2018**, *9*, 947. [CrossRef] [PubMed]

117. Zajac, P.; Schultz-Thater, E.; Tornillo, L.; Sadowski, C.; Trella, E.; Mengus, C.; Iezzi, G.; Spagnoli, G.C. MAGE-A Antigens and Cancer Immunotherapy. *Front. Med.* **2017**, *4*, 18. [CrossRef] [PubMed]

118. Turriziani, M.; Fantini, M.; Benvenuto, M. Carcinoembryonic antigen (CEA)-based cancer vaccines: Recent patents and antitumor effects from experimental models to clinical trials. *Recent Pat. Anti-Cancer Drug Discov.* **2012**, *7*, 265–296. [CrossRef]

119. Evdokimova, V.N.; Butterfield, L.H. Alpha-fetoprotein and other tumour-associated antigens for immunotherapy of hepatocellular cancer. *Expert Opin. Biol. Ther.* **2008**, *8*, 325–336. [CrossRef] [PubMed]

120. Westdorp, H.; Sköld, A.E.; Snijer, B.A.; Franik, S.; Mulder, S.F.; Major, P.P.; Foley, R.; Gerritsen, W.R.; de Vries, I.J.M. Immunotherapy for prostate cancer: Lessons from responses to tumor-associated antigens. *Front. Immunol.* **2014**, *5*, 191. [CrossRef]

121. Scheuermann, R.H.; Racila, E. CD19 antigen in leukemia and lymphoma diagnosis and immunotherapy. *Leuk. Lymphoma* **1995**, *18*, 385–397. [CrossRef]

122. Ayoub, N.M.; Al-Shami, K.M.; Yaghan, R.J. Immunotherapy for HER2-positive breast cancer: Recent advances and combination therapeutic approaches. *Breast Cancer (Dove Med Press)* **2019**, *11*, 53–69. [CrossRef]

123. Liu, J.-P.; Chen, W.; Schwarer, A.P.; Li, H. Telomerase in cancer immunotherapy. *Biochim. Biophys. Acta* **2010**, *1805*, 35–42. [CrossRef]

124. Jiang, T.; Shi, T.; Zhang, H.; Hu, J.; Song, Y.; Wei, J.; Ren, S.; Zhou, C. Tumor neoantigens: From basic research to clinical applications. *J. Hematol. Oncol.* **2019**, *12*, 93. [CrossRef] [PubMed]

125. Goydos, J.S.; Elder, E.; Whiteside, T.L.; Finn, O.J.; Lotze, M.T. A phase I trial of a synthetic mucin peptide vaccine. Induction of specific immune reactivity in patients with adenocarcinoma. *J. Surg. Res.* **1996**, *63*, 298–304. [CrossRef]

126. Mukherji, B.; Chakraborty, N.G.; Yamasaki, S.; Okino, T.; Yamase, H.; Sporn, J.R.; Kurtzman, S.K.; Ergin, M.T.; Ozols, J.; Meehan, J. Induction of antigen-specific cytolytic T cells in situ in human melanoma by immunization with synthetic peptide-pulsed autologous antigen presenting cells. *Proc. Natl. Acad. Sci. USA* **1995**, *92*, 8078–8082. [CrossRef] [PubMed]

127. Hsu, F.J.; Benike, C.; Fagnoni, F.; Liles, T.M.; Czerwinski, D.; Taidi, B.; Engleman, E.G.; Levy, R. Vaccination of patients with B-cell lymphoma using autologous antigen-pulsed dendritic cells. *Nat. Med.* **1996**, *2*, 52–58. [CrossRef] [PubMed]
128. Kim, J.W.; Gulley, J.L. Poxviral vectors for cancer immunotherapy. *Expert Opin. Biol. Ther.* **2012**, *12*, 463–478. [CrossRef]
129. Gallo, P.; Dharmapuri, S.; Cipriani, B.; Monaci, P. Adenovirus as vehicle for anticancer genetic immunotherapy. *Gene Ther.* **2005**, *12* (Suppl. 1), S84–S91. [CrossRef]
130. Roland, K.L.; Brenneman, K.E. Salmonella as a vaccine delivery vehicle. *Expert Rev. Vaccines* **2013**, *12*, 1033–1045. [CrossRef]
131. Flickinger, J.C.; Rodeck, U.; Snook, A.E. Listeria monocytogenes as a Vector for Cancer Immunotherapy: Current Understanding and Progress. *Vaccines* **2018**, *6*, 48. [CrossRef]
132. Jahanafrooz, Z.; Baradaran, B.; Mosafer, J.; Hashemzaei, M.; Rezaei, T.; Mokhtarzadeh, A.; Hamblin, M.R. Comparison of DNA and mRNA vaccines against cancer. *Drug Discov. Today* **2019**. [CrossRef]
133. Palucka, K.; Banchereau, J. Dendritic-cell-based therapeutic cancer vaccines. *Immunity* **2013**, *39*, 38–48. [CrossRef]
134. Kantoff, P.W.; Higano, C.S.; Shore, N.D.; Berger, E.R.; Small, E.J.; Penson, D.F.; Redfern, C.H.; Ferrari, A.C.; Dreicer, R.; Sims, R.B.; et al. IMPACT Study Investigators Sipuleucel-T immunotherapy for castration-resistant prostate cancer. *N. Engl. J. Med.* **2010**, *363*, 411–422. [CrossRef]
135. Peng, M.; Mo, Y.; Wang, Y.; Wu, P.; Zhang, Y.; Xiong, F.; Guo, C.; Wu, X.; Li, Y.; Li, X.; et al. Neoantigen vaccine: An emerging tumor immunotherapy. *Mol. Cancer* **2019**, *18*, 128. [CrossRef] [PubMed]
136. Ott, P.A.; Hu, Z.; Keskin, D.B.; Shukla, S.A.; Sun, J.; Bozym, D.J.; Zhang, W.; Luoma, A.; Giobbie-Hurder, A.; Peter, L.; et al. An immunogenic personal neoantigen vaccine for patients with melanoma. *Nature* **2017**, *547*, 217–221. [CrossRef] [PubMed]
137. Keskin, D.B.; Anandappa, A.J.; Sun, J.; Tirosh, I.; Mathewson, N.D.; Li, S.; Oliveira, G.; Giobbie-Hurder, A.; Felt, K.; Gjini, E.; et al. Neoantigen vaccine generates intratumoral T cell responses in phase Ib glioblastoma trial. *Nature* **2019**, *565*, 234–239. [CrossRef] [PubMed]
138. Finn, O.J. The dawn of vaccines for cancer prevention. *Nat. Rev. Immunol.* **2018**, *18*, 183–194. [CrossRef]
139. Chang, M.-H. Hepatitis B virus and cancer prevention. *Recent Results Cancer Res.* **2011**, *188*, 75–84. [CrossRef]
140. Guo, F.; Cofie, L.E.; Berenson, A.B. Cervical cancer incidence in young U.S. females after human papillomavirus vaccine introduction. *Am. J. Prev. Med.* **2018**, *55*, 197–204. [CrossRef]
141. Singh, R.; Bandyopadhyay, D. MUC1: A target molecule for cancer therapy. *Cancer Biol. Ther.* **2007**, *6*, 481–486. [CrossRef]
142. Kimura, T.; McKolanis, J.R.; Dzubinski, L.A.; Islam, K.; Potter, D.M.; Salazar, A.M.; Schoen, R.E.; Finn, O.J. MUC1 vaccine for individuals with advanced adenoma of the colon: A cancer immunoprevention feasibility study. *Cancer Prev. Res.* **2013**, *6*, 18–26. [CrossRef]
143. Lawler, S.E.; Speranza, M.-C.; Cho, C.-F.; Chiocca, E.A. Oncolytic viruses in cancer treatment: A review. *JAMA Oncol.* **2017**, *3*, 841–849. [CrossRef]
144. Alvarez, C.A. NK cells impede glioblastoma virotherapy through NKp30 and NKp46 natural cytotoxicity receptors. *Nat. Med.* **2012**, *18*, 1827. [CrossRef]
145. Yung, W.K.A.; Vence, L.M.; Gomez, C. Delta-24-RGD oncolytic adenovirus elicits anti-glioma immunity in an immunocompetent mouse model. *PloS ONE* **2014**, *9*, e97407.
146. Nguyen, T.; Avci, N.G.; Shin, D.H.; Martinez, N. Tune up in situ autovaccination against solid tumors with oncolytic viruses. *Cancers* **2018**, *10*, 171. [CrossRef]
147. Dock, G. The influence of complicating diseases upon leukaemia. *Am. J. Med Sci.* **1904**, *127*, 563. [CrossRef]
148. Larson, C.; Oronsky, B.; Scicinski, J.; Fanger, G.R.; Stirn, M.; Oronsky, A.; Reid, T.R. Going viral: A review of replication-selective oncolytic adenoviruses. *Oncotarget* **2015**, *6*, 19976–19989. [CrossRef] [PubMed]
149. Andtbacka, R.H.I.; Kaufman, H.L.; Collichio, F.; Amatruda, T.; Senzer, N.; Chesney, J.; Delman, K.A.; Spitler, L.E.; Puzanov, I.; Agarwala, S.S.; et al. Talimogene laherparepvec improves durable response rate in patients with advanced melanoma. *J. Clin. Oncol.* **2015**, *33*, 2780–2788. [CrossRef]
150. Rosenberg, S.A.; Restifo, N.P. Adoptive cell transfer as personalized immunotherapy for human cancer. *Science* **2015**, *348*, 62–68. [CrossRef]
151. Cohen, J.E.; Merims, S.; Frank, S.; Engelstein, R.; Peretz, T.; Lotem, M. Adoptive cell therapy: Past, present and future. *Immunotherapy* **2017**, *9*, 183–196. [CrossRef]

152. Mitchison, N.A. Passive transfer of transplantation immunity. *Nature* **1953**, *171*, 267–268. [CrossRef] [PubMed]

153. Delorme, E.J.; Alexander, P. Treatment of primary fibrosarcoma in the rat with immune lymphocytes. *Lancet* **1964**, *2*, 117–120. [CrossRef]

154. Fefer, A. Immunotherapy and chemotherapy of Moloney sarcoma virus-induced tumors in mice. *Cancer Res.* **1969**, *29*, 2177–2183. [PubMed]

155. Morgan, D.A.; Ruscetti, F.W.; Gallo, R. Selective in vitro growth of T lymphocytes from normal human bone marrows. *Science* **1976**, *193*, 1007–1008. [CrossRef] [PubMed]

156. Donohue, J.H.; Rosenstein, M.; Chang, A.E.; Lotze, M.T.; Robb, R.J.; Rosenberg, S.A. The systemic administration of purified interleukin 2 enhances the ability of sensitized murine lymphocytes to cure a disseminated syngeneic lymphoma. *J. Immunol.* **1984**, *132*, 2123–2128. [PubMed]

157. Muul, L.M.; Spiess, P.J.; Director, E.P.; Rosenberg, S.A. Identification of specific cytolytic immune responses against autologous tumor in humans bearing malignant melanoma. *J. Immunol.* **1987**, *138*, 989–995. [PubMed]

158. Rosenberg, S.A.; Spiess, P.; Lafreniere, R. A new approach to the adoptive immunotherapy of cancer with tumor-infiltrating lymphocytes. *Science* **1986**, *233*, 1318–1321. [CrossRef]

159. Rosenberg, S.A.; Packard, B.S.; Aebersold, P.M.; Solomon, D.; Topalian, S.L.; Toy, S.T.; Simon, P.; Lotze, M.T.; Yang, J.C.; Seipp, C.A. Use of tumor-infiltrating lymphocytes and interleukin-2 in the immunotherapy of patients with metastatic melanoma. A preliminary report. *N. Engl. J. Med.* **1988**, *319*, 1676–1680. [CrossRef]

160. Slavin, S.; Nagler, A.; Naparstek, E.; Kapelushnik, Y.; Aker, M.; Cividalli, G.; Varadi, G.; Kirschbaum, M.; Ackerstein, A.; Samuel, S.; et al. Nonmyeloablative stem cell transplantation and cell therapy as an alternative to conventional bone marrow transplantation with lethal cytoreduction for the treatment of malignant and nonmalignant hematologic diseases. *Blood* **1998**, *91*, 756–763. [CrossRef]

161. Dudley, M.E.; Wunderlich, J.R.; Yang, J.C. A phase I study of nonmyeloablative chemotherapy and adoptive transfer of autologous tumor antigen-specific T lymphocytes in patients with metastatic melanoma. *J. Immunother.* **2002**, *25*, 243. [CrossRef]

162. Dudley, M.E.; Wunderlich, J.R.; Robbins, P.F.; Yang, J.C.; Hwu, P.; Schwartzentruber, D.J.; Topalian, S.L.; Sherry, R.; Restifo, N.P.; Hubicki, A.M.; et al. Cancer regression and autoimmunity in patients after clonal repopulation with antitumor lymphocytes. *Science* **2002**, *298*, 850–854. [CrossRef]

163. Morgan, R.A.; Dudley, M.E.; Wunderlich, J.R.; Hughes, M.S.; Yang, J.C.; Sherry, R.M.; Royal, R.E.; Topalian, S.L.; Kammula, U.S.; Restifo, N.P.; et al. Cancer regression in patients after transfer of genetically engineered lymphocytes. *Science* **2006**, *314*, 126–129. [CrossRef]

164. Johnson, L.A.; Morgan, R.A.; Dudley, M.E.; Cassard, L.; Yang, J.C.; Hughes, M.S.; Kammula, U.S.; Royal, R.E.; Sherry, R.M.; Wunderlich, J.R.; et al. Gene therapy with human and mouse T-cell receptors mediates cancer regression and targets normal tissues expressing cognate antigen. *Blood* **2009**, *114*, 535–546. [CrossRef]

165. Robbins, P.F.; Morgan, R.A.; Feldman, S.A.; Yang, J.C.; Sherry, R.M.; Dudley, M.E.; Wunderlich, J.R.; Nahvi, A.V.; Helman, L.J.; Mackall, C.L.; et al. Tumor regression in patients with metastatic synovial cell sarcoma and melanoma using genetically engineered lymphocytes reactive with NY-ESO-1. *J. Clin. Oncol.* **2011**, *29*, 917–924. [CrossRef]

166. Pule, M.A.; Savoldo, B.; Myers, G.D.; Rossig, C.; Russell, H.V.; Dotti, G.; Huls, M.H.; Liu, E.; Gee, A.P.; Mei, Z.; et al. Virus-specific T cells engineered to coexpress tumor-specific receptors: Persistence and antitumor activity in individuals with neuroblastoma. *Nat. Med.* **2008**, *14*, 1264–1270. [CrossRef]

167. Parkhurst, M.R.; Yang, J.C.; Langan, R.C.; Dudley, M.E.; Nathan, D.-A.N.; Feldman, S.A.; Davis, J.L.; Morgan, R.A.; Merino, M.J.; Sherry, R.M.; et al. T cells targeting carcinoembryonic antigen can mediate regression of metastatic colorectal cancer but induce severe transient colitis. *Mol. Ther.* **2011**, *19*, 620–626. [CrossRef]

168. Robbins, P.F.; Lu, Y.-C.; El-Gamil, M.; Li, Y.F.; Gross, C.; Gartner, J.; Lin, J.C.; Teer, J.K.; Cliften, P.; Tycksen, E.; et al. Mining exomic sequencing data to identify mutated antigens recognized by adoptively transferred tumor-reactive T cells. *Nat. Med.* **2013**, *19*, 747–752. [CrossRef]

169. Tran, E.; Turcotte, S.; Gros, A.; Robbins, P.F.; Lu, Y.-C.; Dudley, M.E.; Wunderlich, J.R.; Somerville, R.P.; Hogan, K.; Hinrichs, C.S.; et al. Cancer immunotherapy based on mutation-specific CD4+ T cells in a patient with epithelial cancer. *Science* **2014**, *344*, 641–645. [CrossRef]

170. Chandran, S.S.; Somerville, R.P.T.; Yang, J.C.; Sherry, R.M.; Klebanoff, C.A.; Goff, S.L.; Wunderlich, J.R.; Danforth, D.N.; Zlott, D.; Paria, B.C.; et al. Treatment of metastatic uveal melanoma with adoptive transfer of tumour-infiltrating lymphocytes: A single-centre, two-stage, single-arm, phase 2 study. *Lancet Oncol.* **2017**, *18*, 792–802. [CrossRef]

171. Lo, W.; Parkhurst, M.; Robbins, P.F.; Tran, E.; Lu, Y.-C.; Jia, L.; Gartner, J.J.; Pasetto, A.; Deniger, D.; Malekzadeh, P.; et al. Immunologic Recognition of a Shared p53 Mutated Neoantigen in a Patient with Metastatic Colorectal Cancer. *Cancer Immunol. Res.* **2019**, *7*, 534–543. [CrossRef]

172. Dudley, M.E.; Rosenberg, S.A. Adoptive cell transfer therapy. *Semin. Oncol.* **2007**, *34*, 524–531. [CrossRef]

173. Hakomori, S. Aberrant glycosylation in tumors and tumor-associated carbohydrate antigens. *Adv. Cancer Res.* **1989**, *52*, 257–331.

174. Werdelin, O.; Meldal, M.; Jensen, T. Processing of glycans on glycoprotein and glycopeptide antigens in antigen-presenting cells. *Proc. Natl. Acad. Sci. USA* **2002**, *99*, 9611–9613. [CrossRef]

175. Garrido, F.; Aptsiauri, N.; Doorduijn, E.M.; Garcia Lora, A.M.; van Hall, T. The urgent need to recover MHC class I in cancers for effective immunotherapy. *Curr. Opin. Immunol.* **2016**, *39*, 44–51. [CrossRef]

176. Gross, G.; Waks, T.; Eshhar, Z. Expression of immunoglobulin-T-cell receptor chimeric molecules as functional receptors with antibody-type specificity. *Proc. Natl. Acad. Sci. USA* **1989**, *86*, 10024–10028. [CrossRef]

177. Stancovski, I.; Schindler, D.G.; Waks, T.; Yarden, Y.; Sela, M.; Eshhar, Z. Targeting of T lymphocytes to Neu/HER2-expressing cells using chimeric single chain Fv receptors. *J. Immunol.* **1993**, *151*, 6577–6582. [PubMed]

178. Hwu, P.; Shafer, G.E.; Treisman, J.; Schindler, D.G.; Gross, G.; Cowherd, R.; Rosenberg, S.A.; Eshhar, Z. Lysis of ovarian cancer cells by human lymphocytes redirected with a chimeric gene composed of an antibody variable region and the Fc receptor gamma chain. *J. Exp. Med.* **1993**, *178*, 361–366. [CrossRef] [PubMed]

179. Krause, A.; Guo, H.F.; Latouche, J.B.; Tan, C.; Cheung, N.K.; Sadelain, M. Antigen-dependent CD28 signaling selectively enhances survival and proliferation in genetically modified activated human primary T lymphocytes. *J. Exp. Med.* **1998**, *188*, 619–626. [CrossRef]

180. Finney, H.M.; Lawson, A.D.; Bebbington, C.R.; Weir, A.N. Chimeric receptors providing both primary and costimulatory signaling in T cells from a single gene product. *J. Immunol.* **1998**, *161*, 2791–2797. [PubMed]

181. Haynes, N.M.; Trapani, J.A.; Teng, M.W.L.; Jackson, J.T.; Cerruti, L.; Jane, S.M.; Kershaw, M.H.; Smyth, M.J.; Darcy, P.K. Single-chain antigen recognition receptors that costimulate potent rejection of established experimental tumors. *Blood* **2002**, *100*, 3155–3163. [CrossRef]

182. Lamers, C.H.J.; Sleijfer, S.; Vulto, A.G.; Kruit, W.H.J.; Kliffen, M.; Debets, R.; Gratama, J.W.; Stoter, G.; Oosterwijk, E. Treatment of metastatic renal cell carcinoma with autologous T-lymphocytes genetically retargeted against carbonic anhydrase IX: First clinical experience. *J. Clin. Oncol.* **2006**, *24*, e20-2. [CrossRef]

183. Kershaw, M.H.; Westwood, J.A.; Parker, L.L.; Wang, G.; Eshhar, Z.; Mavroukakis, S.A.; White, D.E.; Wunderlich, J.R.; Canevari, S.; Rogers-Freezer, L.; et al. A phase I study on adoptive immunotherapy using gene-modified T cells for ovarian cancer. *Clin. Cancer Res.* **2006**, *12*, 6106–6115. [CrossRef]

184. Till, B.G.; Jensen, M.C.; Wang, J.; Chen, E.Y.; Wood, B.L.; Greisman, H.A.; Qian, X.; James, S.E.; Raubitschek, A.; Forman, S.J.; et al. Adoptive immunotherapy for indolent non-Hodgkin lymphoma and mantle cell lymphoma using genetically modified autologous CD20-specific T cells. *Blood* **2008**, *112*, 2261–2271. [CrossRef]

185. Hollyman, D.; Stefanski, J.; Przybylowski, M.; Bartido, S.; Borquez-Ojeda, O.; Taylor, C.; Yeh, R.; Capacio, V.; Olszewska, M.; Hosey, J.; et al. Manufacturing validation of biologically functional T cells targeted to CD19 antigen for autologous adoptive cell therapy. *J. Immunother* **2009**, *32*, 169–180. [CrossRef]

186. Kochenderfer, J.N.; Wilson, W.H.; Janik, J.E.; Dudley, M.E.; Stetler-Stevenson, M.; Feldman, S.A.; Maric, I.; Raffeld, M.; Nathan, D.-A.N.; Lanier, B.J.; et al. Eradication of B-lineage cells and regression of lymphoma in a patient treated with autologous T cells genetically engineered to recognize CD19. *Blood* **2010**, *116*, 4099–4102. [CrossRef]

187. Kalos, M.; Levine, B.L.; Porter, D.L.; Katz, S.; Grupp, S.A.; Bagg, A.; June, C.H. T cells with chimeric antigen receptors have potent antitumor effects and can establish memory in patients with advanced leukemia. *Sci. Transl. Med.* **2011**, *3*, 95ra73. [CrossRef]

188. Porter, D.L.; Levine, B.L.; Kalos, M.; Bagg, A.; June, C.H. Chimeric antigen receptor-modified T cells in chronic lymphoid leukemia. *N. Engl. J. Med.* **2011**, *365*, 725–733. [CrossRef]

189. Milone, M.C.; Fish, J.D.; Carpenito, C.; Carroll, R.G.; Binder, G.K.; Teachey, D.; Samanta, M.; Lakhal, M.; Gloss, B.; Danet-Desnoyers, G.; et al. Chimeric receptors containing CD137 signal transduction domains mediate enhanced survival of T cells and increased antileukemic efficacy in vivo. *Mol. Ther.* **2009**, *17*, 1453–1464. [CrossRef]

190. Shimabukuro-Vornhagen, A.; Gödel, P.; Subklewe, M.; Stemmler, H.J.; Schlößer, H.A.; Schlaak, M.; Kochanek, M.; Böll, B.; von Bergwelt-Baildon, M.S. Cytokine release syndrome. *J. Immunother. Cancer* **2018**, *6*, 56. [CrossRef]

191. Kochenderfer, J.N.; Dudley, M.E.; Feldman, S.A.; Wilson, W.H.; Spaner, D.E.; Maric, I.; Stetler-Stevenson, M.; Phan, G.Q.; Hughes, M.S.; Sherry, R.M.; et al. B-cell depletion and remissions of malignancy along with cytokine-associated toxicity in a clinical trial of anti-CD19 chimeric-antigen-receptor-transduced T cells. *Blood* **2012**, *119*, 2709–2720. [CrossRef]

192. Kochenderfer, J.N.; Dudley, M.E.; Kassim, S.H.; Somerville, R.P.T.; Carpenter, R.O.; Stetler-Stevenson, M.; Yang, J.C.; Phan, G.Q.; Hughes, M.S.; Sherry, R.M.; et al. Chemotherapy-refractory diffuse large B-cell lymphoma and indolent B-cell malignancies can be effectively treated with autologous T cells expressing an anti-CD19 chimeric antigen receptor. *J. Clin. Oncol.* **2015**, *33*, 540–549. [CrossRef]

193. Maus, M.V.; Grupp, S.A.; Porter, D.L.; June, C.H. Antibody-modified T cells: CARs take the front seat for hematologic malignancies. *Blood* **2014**, *123*, 2625–2635. [CrossRef]

194. Grupp, S.A.; Kalos, M.; Barrett, D.; Aplenc, R.; Porter, D.L.; Rheingold, S.R.; Teachey, D.T.; Chew, A.; Hauck, B.; Wright, J.F.; et al. Chimeric antigen receptor-modified T cells for acute lymphoid leukemia. *N. Engl. J. Med.* **2013**, *368*, 1509–1518. [CrossRef]

195. Lee, D.W.; Gardner, R.; Porter, D.L.; Louis, C.U.; Ahmed, N.; Jensen, M.; Grupp, S.A.; Mackall, C.L. Current concepts in the diagnosis and management of cytokine release syndrome. *Blood* **2014**, *124*, 188–195. [CrossRef]

196. Yeku, O.; Li, X.; Brentjens, R.J. Adoptive T-Cell Therapy for Solid Tumors. *Am. Soc. Clin. Oncol. Educ. Book* **2017**, *37*, 193–204. [CrossRef] [PubMed]

197. Baybutt, T.R.; Flickinger, J.C.; Caparosa, E.M.; Snook, A.E. Advances in Chimeric Antigen Receptor T-Cell Therapies for Solid Tumors. *Clin. Pharmacol. Ther.* **2019**, *105*, 71–78. [CrossRef] [PubMed]

198. Mukherjee, S. Genomics-Guided Immunotherapy for Precision Medicine in Cancer. *Cancer Biother. Radiopharm.* **2019**. [CrossRef]

Review

Botulinum Neurotoxins and Cancer—A Review of the Literature

Shivam O. Mittal [1] and Bahman Jabbari [2,*]

[1]   Head, Section for Parkinson's Disease and Movement Disorders, Cleveland Clinic Abu Dhabi, Abu Dhabi 112412, UAE; shivamommittal@gmail.com
[2]   Department of Neurology, Yale University School of Medicine, New Haven, CT 06519, USA
*   Correspondence: bahman.jabbari@yale.edu

Received: 10 December 2019; Accepted: 1 January 2020; Published: 5 January 2020

Abstract: Botulinum neurotoxins (BoNT) possess an analgesic effect through several mechanisms including an inhibition of acetylcholine release from the neuromuscular junction as well as an inhibition of specific pain transmitters and mediators. Animal studies have shown that a peripheral injection of BoNTs impairs the release of major pain transmitters such as substance P, calcitonin gene related peptide (CGRP) and glutamate from peripheral nerve endings as well as peripheral and central neurons (dorsal root ganglia and spinal cord). These effects lead to pain relief via the reduction of peripheral and central sensitization both of which reflect important mechanisms of pain chronicity. This review provides updated information about the effect of botulinum toxin injection on local pain caused by cancer, painful muscle spasms from a remote cancer, and pain at the site of cancer surgery and radiation. The data from the literature suggests that the local injection of BoNTs improves muscle spasms caused by cancerous mass lesions and alleviates the post-operative neuropathic pain at the site of surgery and radiation. It also helps repair the parotid damage (fistula, sialocele) caused by facial surgery and radiation and improves post-parotidectomy gustatory hyperhidrosis. The limited literature that suggests adding botulinum toxins to cell culture slows/halts the growth of certain cancer cells is also reviewed and discussed.

Keywords: botulinum toxin; botulinum neurotoxin; cancer; cancer cells; neuropathic pain; post-surgical pain; parotid gland; submaxillary gland; gustatory hyperhidrosis; sialocele; parotid fistula

Key Contribution: This review demonstrates that local injection of botulinum toxins can alleviate neuropathic pain experienced at the site of surgery and radiation in cancer patients. Local injection of botulinum neurotoxins can also heal fistula and sialocele in the parotid gland of cancer patients damaged by surgical trauma and radiation.

---

1. Introduction

Currently, there are vast indications for the use of botulinum neurotoxins (BoNT) type A and B in clinical medicine. Their specific inhibitory action on cholinergic synapses makes them desirable for the treatment of several hyperkinetic movement disorders as well as symptoms caused by glandular hyperactivity (sialorrhea and hyperhidrosis) and bladder dysfunction [1]. Disease-oriented reviews indicate that these agents are frequently used for the treatment of spasticity in several common disease conditions such as stroke, cerebral palsy, multiple sclerosis, cerebral, and spinal cord injury [2]. The efficacy of BoNT therapy in migraine headaches, predicted by early investigators [3], has been proven via two large, multicenter clinical trials leading to the approval of onabotulinumtoxinA for the treatment of chronic migraine [4]. Animal and human studies have shown that the local injection of botulinum toxins has an analgesic effect and can relieve several forms of neuropathic pain [5–7].

The data indicate an analgesic activity for BoNTs in a wide range of pain disorders that include both neuropathic and non-neuropathic pain.

In recent years, several publications have drawn attention to the utility of BoNT injections in cancer-related pain syndromes arising either by direct pressure from a neoplastic mass or from neuropathic pain at the site of cancer surgery or radiation [8]. Aside from pain, BoNT injection into parotid or submaxillary glands has been shown to reduce symptoms such as sialorrhea resulting from gland injury as well as healing surgical complications such as fistula and sialocele [9]. BoNT injections have been reported to relieve gustatory hyperhidrosis resulting from parotid and oral surgery in cancer patients [10]. The limited literature also suggests that adding BoNT to the culture of cancer cell lines slows growth and mitotic activity of certain cancer cells and promotes apoptosis [11].

This review is based on a literature search using the search engines of Pub Med, Ovid embrace, and Google Scholar from 1989 to 1 September 2019. The terms botulinum toxin, botulinum neurotoxin, onabotulinumtoxinA, incobotulinumtoxinA, abobotulinumtoxinA, and rimabotulinumtoxinB were crossed with cancer pain, postsurgical cancer pain, post-radiation cancer pain, salivary glands, sialorrhea, gustatory sweating, cancer cells, and cancer cell line. Book chapters that were written over the past 10 years focusing on the subject of botulinum toxin therapy in cancer patients were also reviewed. The inclusion criteria encompassed all articles found via those three afore-mentioned search engines using the above-mentioned search words. Inclusion required that the manuscript's abstract contain both words cancer (or neoplasm) and botulinum toxin (or botulinum neurotoxin) therapy. Manuscripts that did not have both cancer and botulinum toxin therapy noted in the abstract were excluded. Manuscripts with benign mass lesions were also excluded.

2. Results

The search identified 746 manuscripts from which 76 were relevant to the subject of botulinum toxins and cancer (see flow chart in Figure 1). After eliminating 12 duplications (due to an overlap between MedLine and Google Scholar), 64 manuscripts remained for final analysis. The collected data can be classified under 3 categories: (1) The role of botulinum toxins in post-radiation and post-surgical cancer pain; (2) the repairing and healing function of BoNT injections upon parotid gland damaged by radiation or surgery; and (3) the effect of botulinum toxins on cancer cell line, cell growth, and apoptosis.

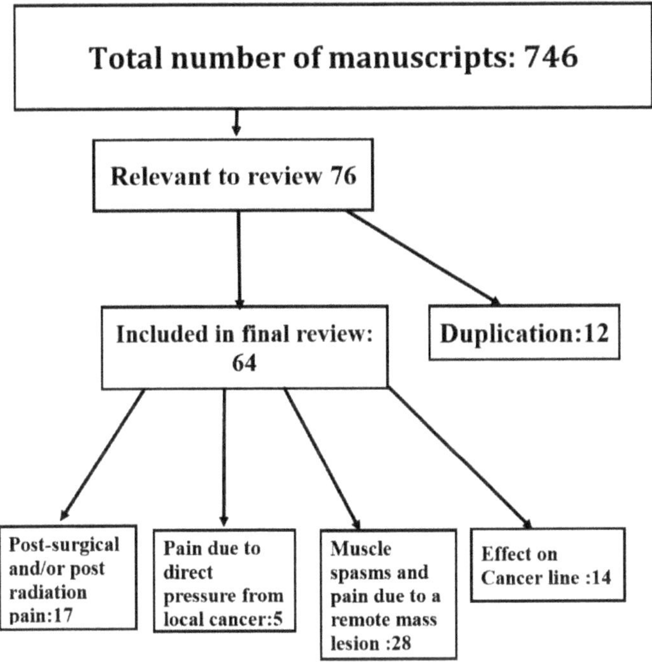

Figure 1. Flow chart of the reviewed manuscripts on cancer and botulinum neurotoxin therapy.

2.1. Botulinum Neurotoxins Therapy for Post-Radiation And/Or Post-Surgical Cancer Pain

This heading includes six prospective clinical trials, four retrospective studies, one double blind placebo-controlled study, and six single case reports (Table 1). A majority of patients had burning and searing pain along the region of fibrosis and keloid formation (neuropathic pain). Some also experienced additional local muscle spasms close to the scarred tissue affecting the neck and shoulder muscles. The most affected muscles were sternocleidomastoid, splenius capitus, trapezius, and levator scapulae. Injections were either subcutaneous (close or at the area of keloid and post-surgical scars, see Figure 2) or both subcutaneous and intramuscular.

Table 1. Published studies on the effect of botulinum neurotoxins (BoNT) on local pain resulting from radiation and/or surgery *.

Authors	Pts Study	Toxin	Dose Units	Treatment	Location of Cancer	Primary Outcome	Result
Van Daele et al., 2002 [12]	6 Retro	OnaA	20–25	Radiation chemotherapy	Head and neck	Pain VAS	Complete pain relief in four of six patients. Significant improvement of quality of life using SF36, EQ-5D scales
Layeeque et al., 2004 [13]	48 Pro	OnaA	100	Mastectomy, assessed for pain after expander placement	Breast	(1) Pain assessed by VAS (2) Narcotic use	Less pain in BoNT group ($p < 0.00001$) Less narcotic use in BoNT group ($p < 0.0001$)
Vasan et al., 2004 [14]	16 Pro	AboA	100 to 320	Surgery	Head and neck	Pain (VAS- days 3 and 4 weeks), global. Quality of life	Significant pain reduction ($p = 0.05$); Quality of life improved ($p = 0.7$)
Wittekindt et al., 2006 [15]	23 Pro	OnaA	60–120 160–240	Radiation; surgery	Head and neck	Pain VAS: 28 weeks	Significant reduction of pain (<0.05)
Hartl et al., 2008 [16]	19 Pro	OnaA AboA	50 250	Chemotherapy; radiation	Head and neck	Pain: (VAS) Function: At 4 weeks	Improved pain ($p = 0.02$) Function ($p = 0.04$)
Stubblefield et al., 2008 [17]	23 Retro	OnaA	25–200	Radiation; surgery	Head, neck, breast	Pain (VAS)	Pain improved in Improved in 85% of patients
Mittal et al., 2012 [18]	7 Retro	OnaA	100	Radiation; surgery	Head, neck, breast	Pain (VAS) PGIC At 4 weeks	VAS: Six of seven patients improved: $p < 0.05$ PGIC: Six of seven, very satisfied QoL: Six of seven improved ($p < 0.05$)
Bach et al., 2012 [19]	9 Pro	OnaA	100–400	Radiation and surgery	Head and neck	Pain (VAS) and FDSNP at 4 weeks	Both pain and FDSNP improved ($p < 0.01$)
Rostami et al., 2014 [20]	12 Pro	IncoA	100	Radiation and surgery	Head neck breast	Pain (VAS) and PGIC at week 6	VAS improved ($p < 0.05$) PGIC: Very satisfied QoL improved in 38% of patients ($p < 0.05$)
De Groef et al. 2018 [21]	50 DBPC	onaA	100	Surgery	Breast	Pain measured by VAS	Pain reduction 60% in the BoNT and 40% in saline group (statistically ns)
Mailly et al., 2019 [22]	16 Retro	incoA aboA	20 40	Radiation and surgery	Head and neck	Pain (VAS)	VAS improved $p < 0.01$

* Case reports are not included. onaA: OnabotulinumtoxinA; aboA: AbobotulinumtoxinA; incoA: IncobotulinumtoxinA; VAS: Visual Analogue Scale; PGIC: Patient Global Impression of Change; FDSNP: Functional Disability Scale for Neck Pain; Pro: Prospective; Retro: Retrospective; DBPC: double-blind, placebo-controlled; QoL: Quality of Life.

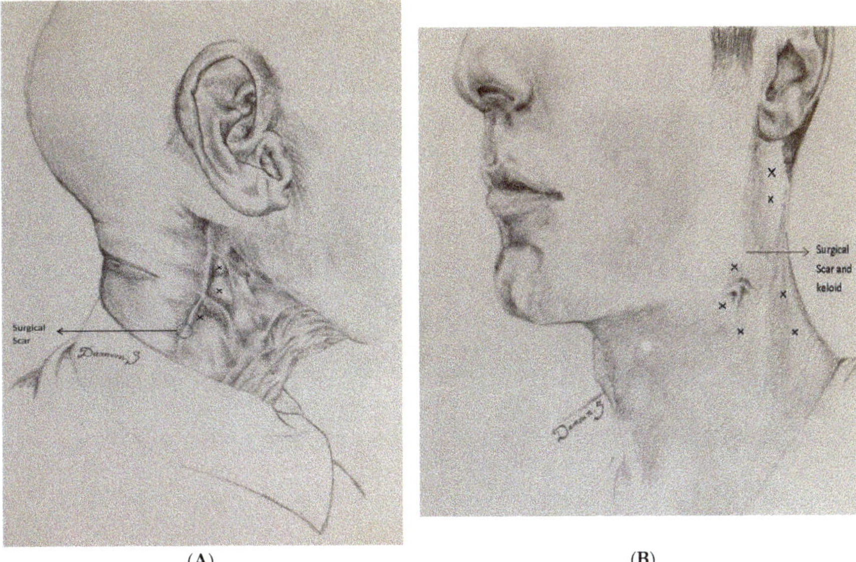

(**A**)   (**B**)

Figure 2. Post-surgical and post-radiation pain treated with BoNT. Example of two patients. From Jabbari B. Botulinum Toxin Treatment in Pain Disorders. Springer, New York 2015. Printed with permission from the publisher. (**A**) A 47-year-old man had undergone right neck dissection and radiotherapy for cancer of the tongue and cervical adenopathy 6 years prior to visiting the Yale clinic. A year following surgery and radiotherapy, severe pain (VAS 9–10, both sharp and deep) developed over the right side of the neck which was mostly felt below the mandible and anterior to the angle of the jaw. Injecting onabotulinumtoxinA into the areas designated by X on the figure, (30, 30, and 20 units) reduced the pain significantly (VAS 1) within a week after injection. He remained responsive and satisfied (assessed by PGIC) receiving injections every 4–6 months over 7 years of follow-up. (**B**) A 48-year old man with squamous cell carcinoma of piriform sinus had supraglottic laryngectomy. Two years following neck dissection and radiotherapy, he developed severe pain (VAS 9) over the left side of the neck. The pain was deep as well as sharp and superficial. Injection of onabotulinumtoxinA, 20 units into each superficial pain region (Xs around the jaw) and 30 units into nearby posteriorly located muscles (splenius and trapezius) designated by X reduced the pain to VAS 0–1 level. The total dose was 200 units. The patient enjoyed pain relief with repeat injections over the 3 years of follow-up. Drawings courtesy of Damoun Safarpour M.D.

A total of 10 of 11 studies used a standardized scale for pain measurement (Visual Analogue Scale: VAS) which in 9 of 10 demonstrated statistically significant improvement of local pain at 4–8 post-injection weeks compared to baseline ($p < 0.05$) (Table 1). Two studies included patient global impression of change (PGIC) in the evaluation, using a 7-grade scale ranging from "very unsatisfied" to "very satisfied". In both studies, patients expressed significant satisfaction with the results [18,20]. Four studies used a scale for evaluating quality of life. Three of 4 demonstrated significant improvement of quality of life after BoNT-A (onaA and incoA) injection therapy [14,18,20]. One study demonstrated significant reduction of daily opioid use after BoNT therapy [13]. One prospective study provided long-term follow up of up to 82 months [20]. Side effects consisted mainly of transient pain at the site of injection(s) and minor local bleeding. None of the 229 patients, reported in Table 1, demonstrated any serious side effect following BoNT injections.

Clinical data from case reports includes six single case reports. One publication reported a 50 year-old man with adenocarcinoma of the soft palate, who following radiotherapy, developed trismus and myokymia of the masseter muscles. Trismus and myokymia improved after injection of 25 units of onaA into each masseter muscle [23]. Two other manuscripts described improvement of central neuropathic pain in association with a mass lesion. One described a 55 year-women who developed severe burning pain and allodynia in the distribution of T1 dermatomes bilaterally following partial resection of an angioma at the C7–C8 region. Subcutaneous injection of onaA at 25 sites into T1 dermatomes (100 units on each sides) resulted in a marked reduction of neuropathic pain and allodynia. This effect was sustained with repeated injections over a follow up period of three years [24]. A similar experience with central pain was reported by Nam et al. [25], in a 62-year old man who had developed severe allodynia and neuropathic pain over the posterior aspect of the left thigh contralateral to a frontal lobe malignant brain tumor. A subcutaneous injection of onaA with a total dose of 100 units at 16 sites substantially improved the patient's neuropathic pain and allodynia over the affected region. In another patient, radiation of a left submandibular chondrosarcoma resulted in hyperactivity of the spinal accessory nerve and gradual painful hypertrophy of the left trapezius muscle. An injection of 90 units of onaA resulted in a substantial reduction of left shoulder pain and diminished the involuntary myokymic movements of the left trapezius muscle [26]. Boukovalas et al. [27] reported a patient with squamous cell carcinoma of the anterior mandible who, following mandibulectomy, bilateral neck dissection, and radiotherapy, gradually developed pain and tightness of the sternocleidomastoid and platysmal muscles associated with Raynaud phenomenon of the lower face. Injection of botulinum toxin (type and dose not mentioned) into the above-mentioned muscles improved painful muscle tightness and reduced the Raynaud phenomena. Schuler et al. [28] described a 47-year old female who, at the scarred skin site of resected melanoma, developed severe neuropathic pain. Injection of onaA, 50 units in a grid-like pattern (injection sites were 1.5 cm apart), resulted in 50% reduction of pain four weeks after BoNT injection.

The duration of action of BoNT injections for pain relief in the above-mentioned studies was 3–6 months (mean 3.9 month). In most studies, the follow up was short term, not exceeding 6–12 months. In some cases, however, patients were followed-up for years with repeated injections. Two patients described in Figure 1 were followed-up for 3 and 7 years (see figure legend). Nine of 11 studies reported no side effects. One study reported increased pain for a few days at the site of injection in one patient, which was followed by baseline pain improvement [14]. One study reported the occurrence of a diffuse maculo-papular rash in one patient 2–3 days after the botulinum neurotoxin injection after which the rash disappeared over a month [20].

2.2. Botulinum Neurotoxins Therapy for Post-Radiation or Postsurgical Damage to Parotid Gland

This category includes six prospective clinical trials, 10 retrospective studies, and 12 single case reports (Table 2). All prospective studies are open label. Botulinum toxin treatment was used for the remedy of post-parotidectomy complications such as gustatory hyperhidrosis (GH), post-parotidectomy sialorrhea, fistula, and sialocele formation.

Table 2. BoNT therapy for post-parotidectomy gustatory hyperhidrosis, fistula, sialocele formation, and for post-parotidectomy sialorrhea.

Authors	Design	Pts #	Clinical Problem	Injection Site	Toxin and Dose	Result
Laskawi et al., 2013 [29]	R	10	Post-parotidectomy fistula	Parotid gland	OnaA 30–50 units	Treated within 6 weeks of surgery: Fistulas healed in 9 of 10 patients
Marchese-Ragona et al., 2006 [30]	R	3	Post-parotidectomy fistula	Parotid gland	OnaA 15–20 units	Complete healing of fistula with follow ups 12,18, and 14 months
Nolte et al., 2004 [31]	P	20	Gustatory sweating after parotidectomy	Facial skin	OnaA 3 units/cm	Complete loss of sweating for 12 months

Table 2. *Cont.*

Authors	Design	Pts #	Clinical Problem	Injection Site	Toxin and Dose	Result
Kuttner et al., 2001 [32]	R	8	GH after parotidectomy	Face	BoNT-A 0.5 units/cm	Stopped facial sweating within one week
Vargas et al., 2000 [33]	P	4	Post-parotidectomy sialocele-pain	Parotid gland	OnaA 30–50 units	Total resolution in 4 weeks in all patients
Steffen et al., 2014 [34]	R	25	Head and neck cancer FHS: (19) Fistula (6)	Parotid gland	OnaA and incoA: Par: 30 U SM: 20 U	FHS: 11 of 19 improved. Fistula: 4 of 6 improved
Machese et al., 2008 [35]	R	8	Head and neck cancer sialorrhea: 6, fistula: 1, and sialocele: 1	Parotid gland	AboA: 100 U	Fistulas healed. Sialorrhea stopped
Eckardt et al., 2003 [36]	R	33	GH after parotidectomy	Face	OnaA 16 to 80 units	Facial sweating disappeared within a week after injections
Cantarella and Barbieri [37]	R	7	GH after parotidectomy	Face	RimaB 2200 units	Cessation of sweating in 6 of 7 patients 4 weeks after injection
Matos Dias et al., 2008 [38]	R	10	GH after parotidectomy	Face	Ona-A 38 units	Sweating stopped
Hatrl et al., 2008 [39]	R	7	GH after parotidectomy	Face	BoNT-A	Sweating and quality of life improved
Pomprasit et al., 2007 [40]	P	9	GH after Parotidectomy	Face	Ona-A 10.6 units	Sweating stopped in 5 and reduced in 4
Cavalot et al., 2000 [41]	P	40	GH after parotidectomy	Face	Ona-A, 2.5/cm^2	100% response in severe group, 72% response in moderate group
Von Lindern et al., 2000 [42]	R	7	GH after parotidectomy	Face	Ona-A	Sweating stopped after BoNT injection
Laccourreye et al., 1998 [43]	P	14	GH after parotidectomy	Face	Ona-A	All showed total cessation of sweating
Bjerkhoel et al., 1997 [44]	P	15	GH after parotidectomy	Face	Ona-A	Total cessation of facial sweating in 13 patients

Case reports are not included for salivary gland problems related to cancer surgery or cancer irradiation. R: Retrospective; P: Prospective; onaA: OnabotulinumtoxinA; incoA: IncobotulinumtoxinA; aboA: AbobotulinumtoxinA; FHS: Functional hypersalivation; Par: Parotid, SM: Submandibular.

The positive information of these studies has been supported by several case reports [45–56]. Among these 12 case reports, six described healing of post-parotidectomy fistula and four reported on healing of sialocele. One case reported improvement of gustatory hyperhidrosis and another one reported improvement of post-parotidectomy sialorrhea. A total of 11 studies had used type-A and one had used type B toxin.

In these 11 studies, no serious side effects were reported. One study reported a patient on anticoagulation in whom a small hematoma developed at the site of injection [34]. One study reported dry mouth as the only side effect [38]. One study cited non-specified, minor issues limited to the site of injection [41]. One study mentioned mild transient weakness of the upper lip in two patients [43] and another study described transient weakness of orbicularis oris muscle in one patient [44].

The clinical studies cited above investigating the analgesic effect of BoNTs in patients after surgery or radiation therapy and BoNT's healing effect on parotid glands injured by surgery or radiation strongly suggest the efficacy of BoNTs in cancer patients affected by surgical and radiation side effects. All three type-A FDA approved BoNTs seem to have analgesic effect in post-surgical and post-radiation pain. In case of parotid injury, at least one study (Table 2) suggests that type B is also effective. Although anecdotal observations have demonstrated safety over 3 to 7 years of treatment (cases presented in Figure 1), the long term safety of BoNT therapy in cancer patients needs to be further investigated through controlled, prospective clinical trials.

2.3. The Effects of Botulinum Neurotoxins Injections on Malignant Tumors and Cancer Cell Line

This category includes 14 studies. In three studies, investigators injected BoNT into a malignant tumor and demonstrated cellular apoptosis and reduction of tumor size [57–59] (Table 3). In another six studies, adding BoNT-A to cancer cell cultures reduced cell growth, induced apoptosis, and inhibited mitosis in various cancer cell lines: Prostate, breast, colon, and pancreatic tumors [60–65]. In one study, transfection of insulin secreting cells by BoNT-A reduced insulin secretion, suggesting a potential for treatment of insulinomas [66]. In another study, the addition of BoNT-A to Her2 positive breast cancer cell line increased Herceptin efficacy [67]. In one study, authors reported no effect on prostate tumor growth and LNCaP and PC3 cancer cells after exposure to BoNT [68]. In one study, increased tumor oxygenation after the injection of BoNT-A into hepatic sarcoma and fibrosarcoma suggested that the BoNT injection potentially made these tumors more susceptible to chemotherapy [69]. In another study, the injection of onaA into one side of cancerous human prostate increased apoptosis on the injected side (compare to saline injected into other side) [70].

Table 3. In vivo and in vitro effects of BoNT injection on malignant tumors and cancer cell lines.

Authors	Study Type	Type of Cells or Tissue	Study Design	Results
Vezdrevanis 2011 [57]	In vivo	Prostatic cancer	Injected BoNT into prostate	Tumor size reduction
Ulloa et al., 2015 [58]	In vivo	Glioblastoma cells	Cells with or without transfection by BoNT-C1 injected into mice striatum	By BoNT-C1 blocks the growth of Glioblastoma cells via blocking Syntaxin1
He et al., 2016 [59]	In vivo	Mice with pancreatic tumor	Injected onaA or saline into tumor	Reduced tumor size; increased apoptosis
Karsenty et al., 2009 [60]	In vitro	Prostate LNCaP and PC-3 cell lines	LNCaP and PC-3 cell lines were exposed to onaA	OnaA inhibited LNCasP cell proliferation; had no effect on PC-3 cell
Nam et al., 2012 [61]	In vitro	Breast and colorectal cancer	PLC-γl-transformed cells were exposed to BoNT-A (difficile)	Caused apoptosis and mitotic inhibition
Proietti et al., 2012 [62]	In vitro	Prostate LNCaP and PC-3 cell lines	Prostate cancer cell lines were exposed to incoA	Tumor cell growth slowed down probably due to toxin effect on SV2 receptors
Bandala et al., 2013 [63]	In vitro	Breast T47D cancer cells	Breast T47D cancer cells were exposed to diverse dilutions of BoNT	BoNT via caspase 3, slow down the growth of T47d cells and caused apoptosis
Bandala et al., 2015 [64]	In vitro	Breast cancer cell line	Added BoNT-A to breast cancer cell line	BoNT-A diminished SV2 protein on the surface of breast cancer cells
Rust et al., 2016 [65]	In vitro	Human neuroblastoma cells	Added BoNT-C to human neuroblastoma cell culture	Apoptosis of neuroblastoma cells
Huang et al., 1998 [66]	Invitro	Insulin secreting HIT-T15 cells	Insulin secreting cells were transfected by BoNT-A	Marked reduction of insulin secretion- potential to treat insulinoma
Hajighasemlou et al., 2015 [67]	In vitro	Her2 positive breast cancer cell line	Assessed the effect of BoNT-A on Her2 positive cells responsive to Herceptin	Herceptin efficacy significantly improved
Cheng et al., 2013 [68]	In vitro and in vivo	Prostate cancer cell line in Mice	LNCaP and PC3 cancer cells were exposed to 1 to 10 units of onaA	No effect on tumor growth in LNCaP and PC3 cancer cells
Ansiaux et al., 2006 [69]	In vivo	Fibrosarcoma, hepatosarcoma	BoNT-A injected into the tumor	Increased oxygenation of the tumor and made it more susceptible to chemo and radiotherapy
Coarfa et al., 2017 [70]	In vivo	Prostate of 250 nude mice Four human cancerous prostates	Effect of onaA versus saline injection into cancer cells implanted into rodent's prostate Assessed the effect of onaA versus saline injection in cancerous prostate before prostatectomy	Increased apoptosis; slowed cancer progression Increased apoptosis in ona-A injected side of prostate

OnaA: OnabotulinumtoxinA (Botox). IncoA: IncobotulinumtoxinA (Xeomin).

3. Discussion

Botulinum neurotoxins exert their analgesic effect through two known mechanisms. The inhibitory effect of the BoNTs upon the release of acetylcholine at the neuromuscular junction is mostly responsible for the relief of pain caused by muscle spasms. In the case of neuropathic pain, it is currently believed that the analgesic effect of botulinum injections predominantly results from inhibition of pain neurotransmitters both at peripheral and at central sensory levels [5,6,71,72]. The peripheral injection of botulinum toxin-A into the muscle or close to peripheral nerve endings reduces the release of calcitonin gene related peptide, a major pain transmitter from trigeminal ganglion [73]. Direct exposure of dorsal root ganglia to botulinum toxin-A significantly reduces the thermal sensitivity in the animal model of thermal pain [74]. In the formalin pain model, injection botulinum toxin B into the rat's paw reduced substance P release from ipsilateral sensory spinal neurons and prevented spinal sensory neuron activation (c-Fos) which occurred after formalin injection [75]. Injection of botulinum toxins into mice hind paw reduces glutamate release from spinal sensory neurons [76]. Intra-articular injection of botulinum toxin in animal models of pain reduces upregulation of transient receptor potential cation channel subfamily V member 1 (TrpV1), a protein closely associated with pain pathophysiology [77]. A central analgesic function for botulinum toxins has been suggested by studies that have shown the presence of cleaved SNAP-25 in medullary and midbrain sensory regions following the peripheral injection of botulinum toxins [78,79]. Further suggestion for central effects of BoNTs comes from the studies that have demonstrated bilateral improvement of pain sensations after the unilateral injection of botulinum toxin in animal models of diabetic neuropathy and acidic saline injection [80,81]. The analgesic effect of BoNTs results from their direct and indirect effects since patients experience analgesia prior to the muscle relaxation [82].

Pain is a common symptom in cancer patients and when present often impairs the patient's quality of life [83]. Approximately 20–60% of the patients with breast cancer and 30% of the patients with head and neck cancer experience chronic pain localized to the site of radiation or surgery [84]. Post-radiation/surgical pain may be treated with the topical application of a hyaluronic acid, calendula officinalis, trolamine, and lidocaine patch [85,86]. However, sustained relief from pain happens only in 25% of the patients using these remedies [87]. Potent systemic analgesic agents such as opioids provide pain relief in many patients but the development of undesirable side effects including nausea, somnolence, constipation, and addiction complicates their use [88]. Botulinum toxin treatment has two major advantages over these pharmacological remedies. Firstly, the effects of the BoNT-A and B injection lasts 3–6 months. Secondly, the BoNT injection has fewer side effects and is safer when compared to potent analgesic agents. Lack of any serious side effect in the studies cited above supports this statement.

Gustatory hyperhidrosis (Frey syndrome) can be congenital or acquired. Acquired gustatory hyperhidrosis results from injury to the parotid gland or face as well as conditions such as diabetic autonomic neuropathy. Gustatory hyperhidrosis after parotidectomy results from the aberrant innervation of sweat glands from parasympathetic nerves of the parotid region. Facial sweating during chewing and eating is often a cause of social embarrassment. Gustatory hyperhidrosis (GH) is common after parotidectomy and about half of the patients complained of this symptom after surgery [89]. Botulinum neurotoxins via blocking acetylcholine release at autonomic synapses are highly effective in treatment of autonomic dysfunctions such as sialorrhea and hyperhidrosis [90]. In a meta-analysis of literature on Frey syndrome (multiple etiologies) treated with BoNTo, Xie et al. found the effectiveness of BoNT therapy to be present in 98% of the patients [10].

Fistula with sialorrhea and sialocele (entrapped saliva with cyst formation) are two common complications of parotidectomy. Treatment of post-parotidectomy fistula consist of pressure dressing, systemic anticholinergic drugs, suction drain insertion, tympanic neurectomy, and surgery [91]. Overall, the results of the above mentioned surgical and medical strategies in the treatment of parotid fistula is disappointing [92]. Furthermore, side effects of anticholinergic therapy such as memory loss, blurring of vision, dryness of the mouth, and urinary dysfunction are not well tolerated, especially in the elderly. Botulinum toxin injections provides a safe and effective way to suppress sialorrhea and to help heal the fistula.

4. Conclusions

The studies of botulinum toxins in post-surgical and post-radiation pain indicated that the local injection of BoNT improved neuropathic pain and local muscle spasm at/or close to the site of surgery and radiation. The proof of efficacy of botulinum toxin therapy in this form of cancer-related pain, however, awaits the results of blinded and placebo-controlled studies. The same conclusion applies to the use of botulinum neurotoxins in gustatory hyperhidrosis and in the management of post-parotidectomy fistula and sialocele where all open-label studies suggest efficacy. The positive effect of BoNTs on different cancer cell lines and their direct effects upon certain cancerous tumors is encouraging. More studies are necessary to verify these results and if verified to devise a methodology through which BoNT injections can safely be used for the treatment of certain human cancers.

Funding: This research received no external funding.

Conflicts of Interest: The authors declare no conflict of interest.

References

1. Jankovic, J. Botulinum toxin: State of the Art. *Mov. Disord.* **2017**, *32*, 1131–1138. [CrossRef]
2. Jabbari, B. (Ed.) A disease- oriented approach. In *Botulinum Toxin Treatment in Clinical Medicine*; Springer: New York, NY, USA, 2018.
3. Aoki, K.R. Evidence for antinociceptive activity of botulinum toxin type A in pain management. *Headache* **2003**, *43* (Suppl. S1), 9–15. [CrossRef] [PubMed]
4. Dodick, D.W.; Turkel, C.C.; DeGryse, R.E.; Aurora, S.K.; Silberstein, S.D.; Lipton, R.B.; Diener, H.C.; Brin, M.F. OnabotulinumtoxinA for treatment of chronic migraine: Pooled results from the double-blind, randomized, placebo-controlled phases of the PREEMPT clinical program. *Headache* **2010**, *50*, 921–936. [CrossRef] [PubMed]
5. Oh, H.M.; Chung, M.E. Botulinum Toxin for Neuropathic Pain: A Review of the Literature. *Toxins* **2015**, *7*, 3127–3154. [CrossRef] [PubMed]
6. Park, J.; Park, H.J. Botulinum Toxin for the Treatment of Neuropathic Pain. *Toxins* **2017**, *9*, 260. [CrossRef] [PubMed]
7. Safarpour, Y.; Jabbari, B. Botulinum toxin treatment of pain syndromes—An evidence based review. *Toxicon* **2018**, *147*, 120–128. [CrossRef]
8. Shaw, L.; Bazzell, A.F.; Dains, J.E. Botulinum Toxin for Side-Effect Management and Prevention of Surgical Complications in Patients Treated for Head and Neck Cancers and Esophageal Cancer. *J. Adv. Pract. Oncol.* **2019**, *10*, 40–52. [PubMed]
9. Melville, J.C.; Stackowicz, D.J.; Jundt, J.S.; Shum, J.W. Use of Botox (OnabotulinumtoxinA) for the Treatment of Parotid Sialocele and Fistula After Extirpation of Buccal Squamous Cell Carcinoma with Immediate Reconstruction Using Microvascular Free Flap: A Report of 3 Cases. *J. Oral Maxillofac. Surg.* **2016**, *74*, 1678–1686. [CrossRef]
10. Xie, S.; Wang, K.; Xu, T.; Guo, X.S.; Shan, X.F.; Cai, Z.G. Efficacy and safety of botulinum toxin type A for treatment of Frey's syndrome: Evidence from 22 published articles. *Cancer Med.* **2015**, *4*, 1639–1650. [CrossRef]
11. Matak, I.; Lacković, Z. Botulinum neurotoxin type A: Actions beyond SNAP-25? *Toxicology* **2015**, *335*, 79–84. [CrossRef]
12. Van Daele, D.J.; Finnegan, E.M.; Rodnitzky, R.L.; Zhen, W.; McCulloch, T.M.; Hoffman, H.T. Head and neck muscle spasm after radiotherapy: Management with botulinum toxin A injection. *Arch. Otolaryngol. Head Neck Surg.* **2002**, *128*, 956–959. [CrossRef]
13. Layeeque, R.; Hochberg, J.; Siegel, E.; Kunkel, K.; Kepple, J.; Henry-Tillman, R.S.; Dunlap, M.; Seibert, J.; Klimberg, V.S. Botulinum toxin infiltration for pain control after mastectomy and expander reconstruction. *Ann. Surg.* **2004**, *240*, 608–613. [CrossRef] [PubMed]
14. Vasan, C.W.; Liu, W.C.; Klussmann, J.P.; Guntinas-Lichius, O. Botulinum toxin type A for the treatment of chronic neck pain after neck dissection. *Head Neck* **2004**, *26*, 39–45. [CrossRef] [PubMed]

15. Wittekindt, C.; Liu, W.C.; Preuss, S.F.; Guntinas-Lichius, O. Botulinum toxin A for neuropathic pain after neck dissection: A dose-finding study. *Laryngoscope* **2006**, *116*, 1168–1171. [CrossRef] [PubMed]
16. Hartl, D.M.; Cohen, M.; Juliéron, M.; Marandas, P.; Janot, F.; Bourhis, J. Botulinum toxin for radiation-induced facial pain and trismus. *Otolaryngol. Head Neck Surg.* **2008**, *138*, 459–463. [PubMed]
17. Stubblefield, M.D.; Levine, A.; Custodio, C.M.; Fitzpatrick, T. The role of botulinum toxin type A in the radiation fibrosis syndrome: A preliminary report. *Arch. Phys. Med. Rehabil.* **2008**, *89*, 417–421. [CrossRef]
18. Mittal, S.; Machado, D.G.; Jabbari, B. OnabotulinumtoxinA for treatment of focal cancer pain after surgery and/or radiation. *Pain Med.* **2012**, *13*, 1029–1033. [CrossRef]
19. Bach, C.A.; Wagner, I.; Lachiver, X.; Baujat, B.; Chabolle, F. Botulinum toxin in the treatment of post-radiosurgical neck contracture in head and neck cancer: A novel approach. *Eur. Ann. Otorhinolaryngol. Head Neck Dis.* **2012**, *129*, 6–10. [CrossRef]
20. Rostami, R.; Mittal, S.O.; Radmand, R.; Jabbari, B. Incobotulinum Toxin-A Improves Post-Surgical and Post-Radiation Pain in Cancer Patients. *Toxins* **2016**, *8*, 22. [CrossRef]
21. De Groef, A.; Devoogdt, N.; Van Kampen, M.; Nevelsteen, I.; Smeets, A.; Neven, P.; Geraerts, I.; Dams, L.; Van der Gucht, E.; Debeer, P. Effectiveness of Botulinum Toxin A for Persistent Upper Limb Pain After Breast Cancer Treatment: A Double-Blinded Randomized Controlled Trial. *Arch. Phys. Med. Rehabil.* **2018**, *99*, 1342–1351. [CrossRef]
22. Mailly, M.; Benzakin, S.; Chauvin, A.; Brasnu, D.; Ayache, D. Douleurs post-radiques après radiothérapie pour cancer des voies aérodigestives superieures: Traitement par injections de toxine botulique A [Radiation-induced head and neck pain: Management with botulinum toxin a injections]. *Cancer Radiother.* **2019**, *23*, 312–315. [CrossRef] [PubMed]
23. Lou, J.S.; Pleninger, P.; Kurlan, R. Botulinum toxin A is effective in treating trismus associated with postradiation myokymia and muscle spasm. *Mov. Disord.* **1995**, *10*, 680–681. [CrossRef] [PubMed]
24. Jabbari, B.; Maher, N.; Difazio, M.P. Botulinum toxin a improved pain and allodynia in two patients with intracranial pathology. *Pain Med.* **2003**, *4*, 206–210. [CrossRef] [PubMed]
25. Nam, K.E.; Kim, J.S.; Hong, B.Y. Botulinum Toxin Type A Injection for Neuropathic Pain in a Patient With a Brain Tumor: A Case Report. *Ann. Rehabil. Med.* **2017**, *41*, 1088. [CrossRef]
26. Filippakis, A.; Ho, D.T.; Small, J.E.; Small, K.M.; Ensrud, E.R. Radiation-induced painful neurogenic hypertrophy Treated with Botulinum Toxin, A. *J. Clin. Neuromuscul. Dis.* **2018**, *19*, 135–137. [CrossRef]
27. Boukovalas, S.; Mays, A.C.; Selber, J.C. Botulinum Toxin Injection for Lower Face and Oral Cavity Raynaud Phenomenon after Mandibulectomy, Free Fibula Reconstruction, and Radiation Therapy. *Ann. Plast. Surg.* **2019**, *82*, 53–54. [CrossRef]
28. Schuler, A.; Veenstra, J.; Ozog, D. Battling Neuropathic Scar Pain with Botulinum Toxin. *J. Drugs Dermatol.* **2019**, *18*, 937–938.
29. Laskawi, R.; Winterhoff, J.; Köhler, S.; Kottwitz, L.; Matthias, C. Botulinum toxin treatment of salivary fistulas following parotidectomy: Follow-up results. *Oral Maxillofac. Surg.* **2013**, *17*, 281–285. [CrossRef]
30. Marchese-Ragona, R.; Marioni, G.; Restivo, D.A.; Staffieri, A. The role of botulinum toxin in postparotidectomy fistula treatment. A technical note. *Am. J. Otolaryngol.* **2006**, *27*, 221–224. [CrossRef]
31. Nolte, D.; Gollmitzer, I.; Loeffelbein, D.J.; Hölzle, F.; Wolff, K.D. Botulinumtoxin zur Behandlung des gustatorischen Schwitzens. Eine prospektive randomisierte Therapiestudie [Botulinum toxin for treatment of gustatory sweating. A prospective randomized study]. *Mund Kiefer Gesichtschir.* **2004**, *8*, 369–375. [CrossRef]
32. Kuttner, C.; Tröger, M.; Dempf, R.; Eckardt, A. Effektivität von Botulinumtoxin A in der Behandlung des gustatorischen Schwitzens [Effectiveness of botulinum toxin A in the treatment of gustatory sweating]. *Nervenarzt* **2001**, *72*, 787–790. [CrossRef] [PubMed]
33. Vargas, H.; Galati, L.T.; Parnes, S.M. A pilot study evaluating the treatment of postparotidectomy sialoceles with botulinum toxin type A. *Arch. Otolaryngol. Head Neck Surg.* **2000**, *126*, 421–424. [CrossRef] [PubMed]
34. Steffen, A.; Hasselbacher, K.; Heinrichs, S.; Wollenberg, B. Botulinum toxin for salivary disorders in the treatment of head and neck cancer. *Anticancer Res.* **2014**, *34*, 6627–6632. [PubMed]
35. Marchese, M.R.; Almadori, G.; Giorgio, A.; Paludetti, G. Post-surgical role of botulinum toxin-A injection in patients with head and neck cancer: Personal experience. *Acta Otorhinolaryngol. Ital.* **2008**, *28*, 13–16. [PubMed]
36. Eckardt, A.; Kuettner, C. Treatment of gustatory sweating (Frey's syndrome) with botulinum toxin A. *Head Neck* **2003**, *25*, 624–628. [CrossRef] [PubMed]

37. Cantarella, G.; Berlusconi, A.; Mele, V.; Cogiamanian, F.; Barbieri, S. Treatment of Frey's syndrome with botulinum toxin type B. *Otolaryngol. Head Neck Surg.* **2010**, *143*, 214–218. [CrossRef]

38. Martos Díaz, P.; Bances del Castillo, R.; Mancha de la Plata, M.; Gías, L.; Nieto, C.M.; Lee, G.Y.; Guerra, M.M. Clinical results in the management of Frey's syndrome with injections of Botulinum toxin. *Med. Oral Patol. Oral Cir. Bucal* **2008**, *13*, E248–E252.

39. Hartl, D.M.; Julieron, M.; LeRidant, A.M.; Janot, F.; Marandas, P.; Travagli, J.P. Botulinum toxin A for quality of life improvement in post-parotidectomy gustatory sweating (Frey's syndrome). *J. Laryngol. Otol.* **2008**, *122*, 1100–1104. [CrossRef]

40. Pomprasit, M.; Chintrakarn, C. Treatment of Frey's syndrome with botulinum toxin. *J. Med. Assoc. Thail.* **2007**, *90*, 2397–2402.

41. Cavalot, A.L.; Palonta, F.; Preti, G.; Nazionale, G.; Ricci, E.; Staffieri, A.; Di Girolamo, S.; Cortesina, G. Sindrome di Frey post parotidectomia. Trattamento con tossina botulinica di tipo A [Post-parotidectomy Frey's syndrome. Treatment with botulinum toxin type A]. *Acta Otorhinolaryngol. Ital.* **2000**, *20*, 187–191.

42. Von Lindern, J.J.; Niederhagen, B.; Bergé, S.; Hägler, G.; Reich, R.H. Frey syndrome: Treatment with type A botulinum toxin. *Cancer* **2000**, *89*, 1659–1663. [CrossRef]

43. Laccourreye, O.; Muscatelo, L.; Naude, C.; Bonan, B.; Brasnu, D. Botulinum toxin type A for Frey's syndrome: A preliminary prospective study. *Ann. Otol. Rhinol. Laryngol.* **1998**, *107*, 52–55. [CrossRef] [PubMed]

44. Bjerkhoel, A.; Trobbe, O. Frey's syndrome: Treatment with botulinum toxin. *J. Laryngol. Otol.* **1997**, *111*, 839–844. [CrossRef] [PubMed]

45. Ferron, C.; Cernea, S.S.; Almeida, A.R.T.; Cesar, D.V.G. Primary treatment of early fistula of parotid duct with botulinum toxin type A injection. *An. Bras. Dermatol.* **2017**, *92*, 864–866. [CrossRef] [PubMed]

46. Philouze, P.; Vertu, D.; Ceruse, P. Bilateral gustatory sweating in the submandibular region after bilateral neck dissection successfully treated with botulinum toxin. *Br. J. Oral Maxillofac. Surg.* **2014**, *52*, 761–763. [CrossRef] [PubMed]

47. Ihler, F.; Laskawi, R.; Matthias, C.; Rustenbeck, H.H.; Canis, M. Botulinumtoxin A gegen Hypersalivation. Einsatz bei Wundheilung nach Resektion eines Zungenkarzinoms [Botulinum toxin A after microvascular ALT flap in a patient with (corrected) squamous cell carcinoma of the tongue] [published correction appears in HNO. 2012 Oct; 60, 905]. *HNO* **2012**, *60*, 524–527. [CrossRef]

48. Pantel, M.; Volk, G.F.; Guntinas-Lichius, O.; Wittekindt, C. Botulinum toxin type b for the treatment of a sialocele after parotidectomy. *Head Neck* **2013**, *35*, E11–E12. [CrossRef]

49. Steffen, A.; Wollenberg, B.; Schönweiler, R.; Brüggemann, N.; Meyners, T. Drooling nach Strahlentherapie. Botulinumtoxin als erfolgreiches Therapieverfahren [Drooling following radiation. Botulinum toxin as a successful treatment modality]. *HNO* **2011**, *59*, 115–117. [CrossRef]

50. Hill, S.E.; Mortimer, N.J.; Hitchcock, B.; Salmon, P.J. Parotid fistula complicating surgical excision of a basal cell carcinoma: Successful treatment with botulinum toxin type A. *Dermatol. Surg.* **2007**, *33*, 1365–1367. [CrossRef]

51. Kizilay, A.; Aladağ, I.; Ozturan, O. Parotidektomi sonrasi gelişen tükürük fistülünün botulinum toksini ile iyileşmesi [Successful use of botulinum toxin injection in the treatment of salivary fistula following parotidectomy]. *Kulak Burun Bogaz Ihtis Derg.* **2003**, *10*, 78–81.

52. Guntinas-Lichius, O.; Sittel, C. Treatment of postparotidectomy salivary fistula with botulinum toxin. *Ann. Otol. Rhinol. Laryngol.* **2001**, *110*, 1162–1164. [CrossRef] [PubMed]

53. Marchese Ragona, R.; Blotta, P.; Pastore, A.; Tugnoli, V.; Eleopra, R.; De Grandis, D. Management of parotid sialocele with botulinum toxin. *Laryngoscope* **1999**, *109*, 1344–1346. [CrossRef] [PubMed]

54. Báez, A.; Paleari, J.; Durán, M.N.; Rudy, T.; Califano, I.; Barbosa, N.; Casas Parera, I. Síndrome de Frey por submaxilectomía y tratamiento con toxina botulínica [Frey syndrome secondary to submaxillectomy and botulinic treatment]. *Medicina (B Aires)* **2007**, *67*, 478–480.

55. Hatzis, G.P.; Finn, R. Using botox to treat a mohs defect repair complicated by a parotid fistula. *J. Oral Maxillofac. Surg.* **2007**, *65*, 2357–2360. [CrossRef]

56. Birch, J.F.; Varma, S.K.; Narula, A.A. Botulinum toxoid in the management of gustatory sweating (Frey's syndrome) after superficial parotidectomy. *Br. J. Plast. Surg.* **1999**, *52*, 230–231. [CrossRef]

57. Vezdrevanis, K. Prostatic carcinoma shrunk after intraprostatic injection of botulinum toxin. *Urol. J.* **2011**, *8*, 239–241.

58. Ulloa, F.; Gonzàlez-Juncà, A.; Meffre, D.; Barrecheguren, P.J.; Martínez-Mármol, R.; Pazos, I.; Olivé, N.; Cotrufo, T.; Seoane, J.; Soriano, E. Blockade of the SNARE protein syntaxin 1 inhibits glioblastoma tumor growth. *PLoS ONE* **2015**, *10*, e0119707. [CrossRef]

59. He, D.; Manzoni, A.; Florentin, D.; Fisher, W.; Ding, Y.; Lee, M.; Ayala, G. Biologic effect of neurogenesis in pancreatic cancer. *Hum. Pathol.* **2016**, *52*, 182–189. [CrossRef]

60. Karsenty, G.; Rocha, J.; Chevalier, S.; Scarlata, E.; Andrieu, C.; Zouanat, F.Z.; Rocchi, P.; Giusiano, S.; Elzayat, E.A.; Corcos, J. Botulinum toxin type A inhibits the growth of LNCaP human prostate cancer cells in vitro and in vivo. *Prostate* **2009**, *69*, 1143–1150. [CrossRef]

61. Nam, H.J.; Kang, J.K.; Chang, J.S.; Lee, M.S.; Nam, S.T.; Jung, H.W.; Kim, S.K.; Ha, E.M.; Seok, H.; Son, S.W.; et al. Cells transformed by PLC-gamma 1 overexpression are highly sensitive to clostridium difficile toxin A-induced apoptosis and mitotic inhibition. *J. Microbiol. Biotechnol.* **2012**, *22*, 50–57. [CrossRef]

62. Proietti, S.; Nardicchi, V.; Porena, M.; Giannantoni, A. Attività della tossina botulinica A in linee cellulari di cancro prostatico [Botulinum toxin type-A toxin activity on prostate cancer cell lines]. *Urologia* **2012**, *79*, 135–141. [CrossRef] [PubMed]

63. Bandala, C.; Perez-Santos, J.L.; Lara-Padilla, E.; Delgado Lopez, G.; Anaya-Ruiz, M. Effect of botulinum toxin A on proliferation and apoptosis in the T47D breast cancer cell line. *Asian Pac. J. Cancer Prev.* **2013**, *14*, 891–894. [CrossRef] [PubMed]

64. Bandala, C.; Cortés-Algara, A.L.; Mejía-Barradas, C.M.; Ilizaliturri-Flores, I.; Dominguez-Rubio, R.; Bazán-Méndez, C.I.; Floriano-Sánchez, E.; Luna-Arias, J.P.; Anaya-Ruiz, M.; Lara-Padilla, E.; et al. Botulinum neurotoxin type A inhibits synaptic vesicle 2 expression in breast cancer cell lines. *Int. J. Clin. Exp. Pathol.* **2015**, *8*, 8411–8418. [PubMed]

65. Rust, A.; Leese, C.; Binz, T.; Davletov, B. Botulinum neurotoxin type C protease induces apoptosis in differentiated human neuroblastoma cells. *Oncotarget* **2016**, *7*, 33220–33228. [CrossRef]

66. Huang, X.; Wheeler, M.B.; Kang, Y.H.; Sheu, L.; Lukacs, G.L.; Trimble, W.S.; Gaisano, H.Y. Truncated SNAP-25 (1-197), like botulinum neurotoxin A, can inhibit insulin secretion from HIT-T15 insulinoma cells. *Mol. Endocrinol.* **1998**, *12*, 1060–1070. [CrossRef]

67. Hajighasemlou, S.; Alebouyeh, M.; Rastegar, H.; Manzari, M.T.; Mirmoghtadaei, M.; Moayedi, B.; Ahmadzadeh, M.; Parvizpour, F.; Johari, B.; Naeini, M.M.; et al. Preparation of Immunotoxin Herceptin-Botulinum and Killing Effects on Two Breast Cancer Cell Lines. *Asian Pac. J. Cancer Prev.* **2015**, *16*, 5977–5981. [CrossRef]

68. Cheng, Y.T.; Chung, Y.H.; Kang, H.Y.; Tai, M.H.; Chancellor, M.B.; Chuang, Y.C. OnobotulinumtoxinA Has No Effects on Growth of LNCaP and PC3 Human Prostate Cancer Cells. *Low. Urin. Tract Symptoms* **2013**, *5*, 168–172. [CrossRef]

69. Ansiaux, R.; Gallez, B. Use of botulinum toxins in cancer therapy. *Expert Opin. Investig. Drugs* **2007**, *16*, 209–218. [CrossRef]

70. Coarfa, C.; Florentin, D.; Putluri, N.; Ding, Y.; Au, J.; He, D.; Ragheb, A.; Frolov, A.; Michailidis, G.; Lee, M.; et al. Influence of the neural microenvironment on prostate cancer. *Prostate* **2018**, *78*, 128–139. [CrossRef]

71. Matak, I.; Bölcskei, K.; Bach-Rojecky, L.; Helyes, Z. Mechanisms of Botulinum Toxin Type A Action on Pain. *Toxins* **2019**, *11*, 459. [CrossRef]

72. Mittal, S.O.; Safarpour, D.; Jabbari, B. Botulinum Toxin Treatment of Neuropathic Pain. *Semin. Neurol.* **2016**, *36*, 73–83. [CrossRef] [PubMed]

73. Kitamura, Y.; Matsuka, Y.; Spigelman, I.; Ishihara, Y.; Yamamoto, Y.; Sonoyama, W.; Kamioka, H.; Yamashiro, T.; Kuboki, T.; Oguma, K. Botulinum toxin type a (150 kDa) decreases exaggerated neurotransmitter release from trigeminal ganglion neurons and relieves neuropathy behaviors induced by infraorbital nerve constriction. *Neuroscience* **2009**, *159*, 1422–1429. [CrossRef] [PubMed]

74. Omoto, K.; Maruhama, K.; Terayama, R.; Yamamoto, Y.; Matsushita, O.; Sugimoto, T.; Oguma, K.; Matsuka, Y. Cross-Excitation in Peripheral Sensory Ganglia Associated with Pain Transmission. *Toxins* **2015**, *7*, 2906–2917. [CrossRef] [PubMed]

75. Marino, M.J.; Terashima, T.; Steinauer, J.J.; Eddinger, K.A.; Yaksh, T.L.; Xu, Q. Botulinum toxin B in the sensory afferent: Transmitter release, spinal activation, and pain behavior. *Pain* **2014**, *155*, 674–684. [CrossRef] [PubMed]

76. Hong, B.; Yao, L.; Ni, L.; Wang, L.; Hu, X. Antinociceptive effect of botulinum toxin A involves alterations in AMPA receptor expression and glutamate release in spinal dorsal horn neurons. *Neuroscience* **2017**, *357*, 197–207. [CrossRef] [PubMed]

77. Fan, C.; Chu, X.; Wang, L.; Shi, H.; Li, T. Botulinum toxin type A reduces TRPV1 expression in the dorsal root ganglion in rats with adjuvant-arthritis pain. *Toxicon* **2017**, *133*, 116–122. [CrossRef]

78. Restani, L.; Antonucci, F.; Gianfranceschi, L.; Rossi, C.; Rossetto, O.; Caleo, M. Evidence for anterograde transport and transcytosis of botulinum neurotoxin A (BoNT/A). *J. Neurosci.* **2011**, *31*, 15650–15659. [CrossRef]

79. Matak, I.; Bach-Rojecky, L.; Filipović, B.; Lacković, Z. Behavioral and immunohistochemical evidence for central antinociceptive activity of botulinum toxin A. *Neuroscience* **2011**, *186*, 201–207. [CrossRef]

80. Bach-Rojecky, L.; Salković-Petrisić, M.; Lacković, Z. Botulinum toxin type A reduces pain supersensitivity in experimental diabetic neuropathy: Bilateral effect after unilateral injection. *Eur. J. Pharmacol.* **2010**, *633*, 10–14. [CrossRef]

81. Bach-Rojecky, L.; Lacković, Z. Central origin of the antinociceptive action of botulinum toxin type A. *Pharmacol. Biochem. Behav.* **2009**, *94*, 234–238. [CrossRef]

82. Arezzo, J.C. Possible mechanisms for the effects of botulinum toxin on pain. *Clin. J. Pain.* **2002**, *18* (Suppl. S6), S125–S132. [CrossRef]

83. Neufeld, N.J.; Elnahal, S.M.; Alvarez, R.H. Cancer pain: A review of epidemiology, clinical quality and value impact. *Future Oncol.* **2017**, *13*, 833–841. [CrossRef] [PubMed]

84. Schreiber, K.L.; Kehlet, H.; Belfer, I.; Edwards, R.R. Predicting, preventing and managing persistent pain after breast cancer surgery: The importance of psychosocial factors. *Pain Manag.* **2014**, *4*, 445–459. [CrossRef] [PubMed]

85. Fisher, J.; Scott, C.; Stevens, R.; Marconi, B.; Champion, L.; Freedman, G.M.; Asrari, F.; Pilepich, M.V.; Gagnon, J.D.; Wong, G. Randomized phase III study comparing best supportive care to Biafine as a prophylactic agent for radiation-induced skin toxicity for women undergoing breast irradiation: Radiation Therapy Oncology Group [RTOG] 97–13. *Int. J. Radiat. Oncol. Biol. Phys.* **2000**, *48*, 1307–1310. [CrossRef]

86. Chargari, C.; Fromantin, I.; Kirova, Y.M. Importance of local skin treatments during radiotherapy for prevention and treatment of radio-induced epithelitis. *Cancer Radiother.* **2009**, *13*, 259–266. [CrossRef] [PubMed]

87. Kirova, Y.M.; Fromantin, I.; De Rycke, Y.; Fourquet, A.; Morvan, E.; Padiglione, S.; Falcou, M.C.; Campana, F.; Bollet, M.A. Can we decrease the skin reaction in breast cancer patients using hyaluronic acid during radiation therapy-results of phase III randomised trial. *Radiother. Oncol.* **2011**, *100*, 205–209. [CrossRef]

88. Fleming, J.A.; O'Connor, B.D. Use of lidocaine patches for neuropathic pain in a comprehensive cancer center. *Pain Res. Manag.* **2009**, *14*, 381–388. [CrossRef]

89. Wiffen, P.J.; Derry, S.; Moore, R.A. Impact of morphine, fentanyl, oxycodone or codeine on patient consciousness, appetite and thirst when used to treat cancer pain. *Cochrane Database Syst. Rev.* **2014**, *5*, CD011056.

90. Lakraje, A.A.; Moghimi, N.; Jabbari, B. Sialorrhea, anatomy, physiology and treatment with emphasis on the role of botulinum toxins. *Toxins* **2013**, *5*, 1010–1031. [CrossRef]

91. Schindel, J.; Markowicz, H.; Levie, B. Combined surgical-radiological treatment of parotid gland fistulae. *J. Laryngol. Otol.* **1968**, *82*, 867–870. [CrossRef]

92. Lovato, A.; Restivo, D.A.; Ottaviano, G.; Marioni, G.; Marchese-Ragona, R. Botulinum toxin therapy: Functional silencing of salivary disorders. Terapia con tossina botulinica: Silenziamento funzionale dei disordini salivari. *Acta Otorhinolaryngol. Ital.* **2017**, *37*, 168–171. [CrossRef] [PubMed]

Article

Verotoxin-1-Induced ER Stress Triggers Apoptotic or Survival Pathways in Burkitt Lymphoma Cells

Justine Debernardi [1], Catherine Pioche-Durieu [2], Eric Le Cam [3], Joëlle Wiels [1,*,†] and Aude Robert [4,*]

[1] UMR 8126 CNRS, Institut Gustave Roussy, Université Paris-Saclay, 94805 Villejuif, France; justine.debernardi@gmail.com
[2] UMR 7592 CNRS, Institut Jacques Monod, Université Paris Diderot-Paris 7, 75205 Paris CEDEX 13, France; Catherine.DURIEU@ijm.fr
[3] UMR 8200 CNRS, Institut Gustave Roussy, Université Paris-Saclay, 94805 Villejuif, France; eric.lecam@gustaveroussy.fr
[4] INSERM U1279, Institut Gustave Roussy, Université Paris-Saclay, 94805 Villejuif, France
*   Correspondence: joelle.wiels@gustaveroussy.fr (J.W.); aude.robert@gustaveroussy.fr (A.R.)
†   Current adress: UMR 9018, Université Paris-Saclay, Institut Gustave Roussy, 94805 Villejuif, France.

Received: 10 April 2020; Accepted: 7 May 2020; Published: 11 May 2020

Abstract: Shiga toxins (Stxs) expressed by the enterohaemorrhagic *Escherichia coli* and enteric *Shigella dysenteriae 1* pathogens are protein synthesis inhibitors. Stxs have been shown to induce apoptosis via the activation of extrinsic and intrinsic pathways in many cell types (epithelial, endothelial, and B cells) but the link between the protein synthesis inhibition and caspase activation is still unclear. Endoplasmic reticulum (ER) stress induced by the inhibition of protein synthesis may be this missing link. Here, we show that the treatment of Burkitt lymphoma (BL) cells with verotoxin-1 (VT-1 or Stx1) consistently induced the ER stress response by activation of IRE1 and ATF6—two ER stress sensors—followed by increased expression of the transcription factor C/REB homologous protein (CHOP). However, our data suggest that, although ER stress is systematically induced by VT-1 in BL cells, its role in cell death appears to be cell specific and can be the opposite: ER stress may enhance VT-1-induced apoptosis through CHOP or play a protective role through ER-phagy, depending on the cell line. Several engineered Stxs are currently under investigation as potential anti-cancer agents. Our results suggest that a better understanding of the signaling pathways induced by Stxs is needed before using them in the clinic.

Keywords: shiga toxins; Gb3/CD77; apoptosis; ER stress; autophagy; Burkitt lymphoma

Key Contribution: Simultaneous induction of apoptotic and survival pathways by VT-1.

---

1. Introduction

Shiga toxins (Stxs), also known as verotoxins (VTs) or Shiga-like toxins (SLTs), are a family of cytotoxic proteins, structurally and functionally related, that are produced by the enteric pathogens *Shigella dysenteriae* type 1and Stx-producing *Escherichia coli* (STEC). Two major types of Stxs have been described, VT-1 (or Stx1) and VT-2 (or Stx2), which display 56% amino-acid identity. A broad spectrum of human diseases is associated with Stx-producing organisms, ranging from mild watery diarrhea to bloody diarrhea, hemorrhagic colitis, and life threatening hemolytic uremic syndrome (HUS). Infection with Stx producing bacteria continues to be a significant worldwide public health problem. In the absence of a vaccine or effective therapy to treat the disease, prevention and supportive therapies are currently the main tools to fight such contamination [1,2]. An improved understanding of host-cell responses to Stxs would allow the development of more effective treatment. In addition, the identification of intermediate signaling molecules in Stx-induced pathways may constitute therapeutic targets to limit the tissue damage caused by Stxs.

Members of the Stx family consist of a single 32-kDa A-subunit in non-covalent association with five B-subunits. The B-subunit pentamers form a doughnut-shaped structure that recognizes the cell surface receptor. For nearly all Stxs, this receptor is the neutral glycosphingolipid globotriaosylceramide (Gb3) but Stx2e (responsible of the porcine edema disease) preferentially binds to globotetraosylceramide (Gb4) [3,4]. Following Gb3 binding, Stxs are internalized and undergo retrograde transport through the Golgi to the lumen of the endoplasmic reticulum (ER) [5]. In the ER, the A-subunits are proteolytically cleaved into 27 kDa fragments that translocate to the cytoplasm. This active A-subunit is an N-glycosidase which inhibits protein synthesis by removing an adenine from 28S RNA [6].

Deregulation of Gb3 expression has been observed in various malignancies. Gb3 is highly expressed in Burkitt lymphoma (BL) cells [7] and in diverse types of solid tumors, including breast, testicular, and ovarian carcinomas [8–10]. Interestingly, a new imaging technology based on mass spectrometry (MALDI-2-MSI) has been recently developed to study the precise localization of Gb3 containing various fatty acid moieties and of its precursors which should improve our understanding of glycosphingolipid metabolism in cancer cells [11]. The concept of using Stx and its non-active binding subunit, StxB (as a delivery tool), for therapy emerged from cell trafficking experiments performed in the 1990s. Various preclinical studies have been conducted with this toxin. Regression of the tumor mass has been observed in various xenograft models, but the strong cytotoxicity (protein synthesis arrest and induction of apoptosis) of VT-1 can cause significant side effects, especially in normal cells expressing Gb3. Attempts have thus been made to reduce the doses and/or use modified versions of the toxin [12].

Although the cytotoxic pathway induced by these toxins may differ slightly between diverse cell types, it is now clear that they induce cell death through apoptosis. The apoptotic process generally depends on both caspases and molecules stored in mitochondria [13–15] but there are a few exceptions like HeLa cells where the process is mitochondria-independent [16]. We have further explored the signal transduction pathway induced by VT-1 in BL cells and showed that it is a relatively conventional caspase- and mitochondria-dependent pathway, except for the role of BID (a proapoptotic member of the BCL-2 family), since both the full-length and truncated forms of this protein are involved in the process [17–19]. Others have shown that the ER stress response induced by Stxs/VTs in monocytic THP1 cells contributes to caspase 8 activation and thus also takes part in the apoptotic pathway. In these cells, the B-subunit or the holotoxin containing a mutation-induced inactivated A subunit does not induce apoptosis [13]. These data suggest that the delivery of functional holotoxins to the ER is needed to induce apoptosis.

The ER is an organelle with essential functions in eukaryotic cells. It is both the primary site for the correct folding and processing of proteins for secretion or insertion into the cellular membrane and a major intracellular calcium store. The status of protein folding and Ca^{2+} storage is controlled by three major ER stress sensors: the protein IRE1 (inositol requiring enzyme 1), the serine/threonine kinase PERK (PKR-like ER protein kinase), and the transcription factor ATF6 (activating transcription factor 6). These proteins are associated with the chaperone BIP (binding immunoglobulin protein, also called GRP78 or HSPA5). When unfolded proteins accumulate in the ER, BIP dissociates from the sensors, thus allowing their activation [20]. PERK and IRE1 are activated by homo-dimerization and autophosphorylation, whereas ATF6 activation requires translocation to the Golgi and proteolytic cleavage. ER membrane sensors activate signaling pathways that result in transient attenuation of overall protein translation and in activation of the transcription of genes encoding proteins involved in the degradation of misfolded proteins via the ER associated protein degradation pathway (ERAD). This coordinated response is called the unfolded protein response (UPR). Failure to correct protein folding defects or maintain Ca^{2+} homeostasis induces prolonged signaling through the UPR, leading to apoptosis. C/REB homologous protein (CHOP, also known as GADD153) is a key transcription factor involved in UPR which, directly or indirectly, regulates the expression of genes involved in apoptosis, [21–24]. How the UPR switches from the pro-survival to pro-death mode still remains unclear.

In this work, we aimed to better delineate the early stages of the VT-1-induced apoptotic pathway in BL cells to clarify the involvement of ER stress.

2. Results

2.1. ER-Phagy Can Alter VT-1-Induced Apoptosis

We first analyzed by electronic microscopy (EM) the ultrastructure of two different BL cell lines (BL2 and Ramos) treated or not with VT-1. Before treatment, mitochondria of various sizes displayed typical ultrastructure, with the inner membrane projecting into the matrix at crista junctions to form lamellar cristae (BL2, Figure 1(a1,a2); Ramos, Figure 1(a6,a7), black arrowhead). After 6 h of treatment with VT-1, apoptosis was clearly induced in both cell lines; most cells had a fragmented nucleus and a dark cytoplasm containing vacuoles. Furthermore, we noticed marked changes in the mitochondrial ultrastructure; some mitochondria became darker and smaller, suggesting that they were fragmented, and some showed the loss of cristae, with empty spaces (BL2, Figure 1(a3–a5); Ramos, Figure 1(a8–a10), white arrowhead). This aspect of the mitochondria is consistent with membrane permeability and apoptogen factor release.

However, we also observed additional features in Ramos, reminiscent of the morphology of autophagic organelles. At an earlier timepoint of treatment (4 h) and in some cells at 6 h, we observed large vesicles that resembled autophagosomes, except that the delimiting outer membranes were densely studded with ribosomes, suggesting that the membranes were derived from the ER (Figure 1(a12,a13), white arrows, named ring-shaped ER whorls). We also observed an expanded and more peripheral rough ER (Figure 1(a11), black arrow) and more cytoplasmic rough ER extensions (Figure 1(a9), black arrows), which are characteristics of ER-phagy.

We thus evaluated the effects of autophagy inhibitors such as chloroquine (CQ, Figure 1(b1)) and 3-methyladenine (3MA, Figure 1(b2)) on VT-1-induced apoptosis. Neither CQ nor 3MA had any effects on the induction of apoptosis in BL2 cells (26 ± 6% of apoptotic cells with VT-1 and CQ versus 25 ± 3% with VT-1 alone and 41 ± 10% with VT-1 and 3MA versus 43 ± 15% with VT-1 alone). By contrast, treatment of Ramos cells with autophagy inhibitors increased VT-1-induced cytotoxicity (from 49 ± 5% to 64 ± 7% for CQ and from 38 ± 8% to 57 ± 10% for 3MA), suggesting that autophagy, and more precisely ER-phagy, protects these cells from VT-1 induced cell death.

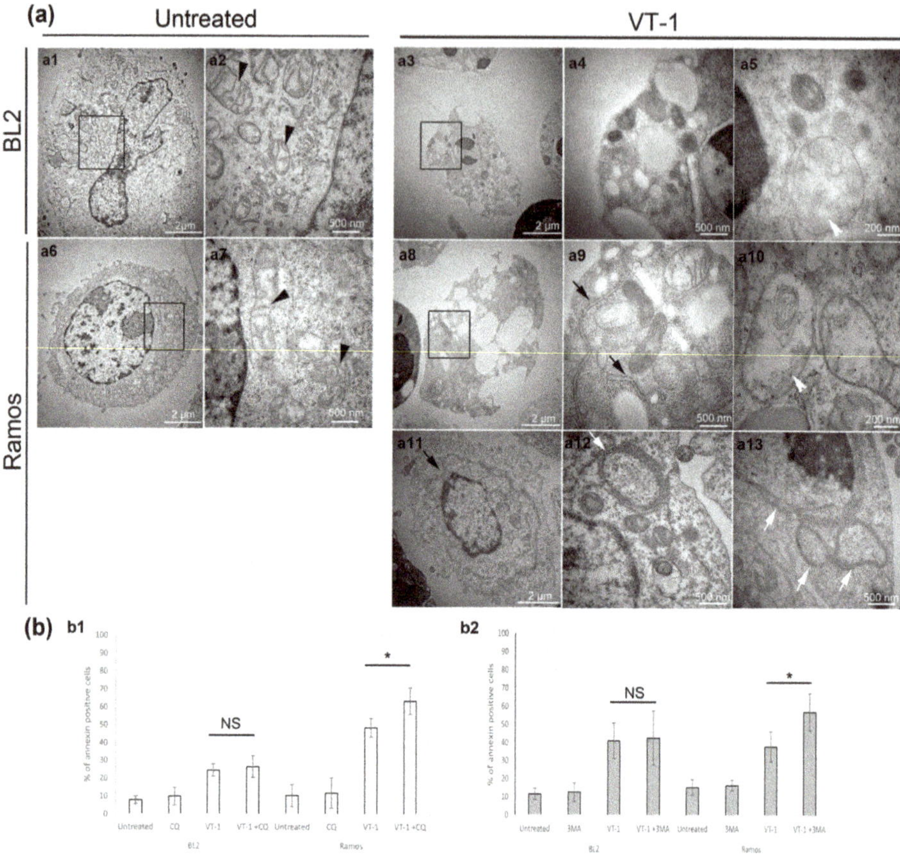

Figure 1. Verotoxin-1 (VT-1) induced protective endoplasmic reticulum (ER)-phagy in certain Burkitt lymphoma (BL) cell lines. (**a**) Effect of VT-1 treatment observed by electron microscopy. (**a1–a5**) BL2 cells; (**a6–a13**) Ramos cells. (**a2,a4,a7,a9**) show magnifications of (**a1,a3,a6,a8**), respectively. Black arrow head: mitochondria with lamellar cristae, white arrow head: mitochondria with the loss of cristae and empty spaces, black arrow: expanded peripheral ER, cytoplasmic ER extension; white arrows: ring-shaped ER whorls. Cells in (**a3–a5**) and (**a8–a10**) were treated with VT-1 for 6 h and for 4 h in (**a11–a13**). (**b**) Effect of chloroquine (CQ) and 3-methyladenine (3MA), two different autophagy inhibitors on VT-1-induced apoptosis. Cells were pretreated with CQ (**b1**) or 3MA (**b2**) for 30 min and subsequently with VT-1 for 6 h. Cells were then analyzed by flow cytometry after annexin V/PI labeling. Statistical test: Mann Whitney, * $p < 0.05$. NS, non significant. Scale bars (**a1,a3,a6,a8,a11**), 2 µm; (**a2,a4,a7,a9,a12,a13**), 500 nm; (**a5,a10**), 200 nm.

2.2. VT-1 Induces the ER Stress Response in Burkitt lymphoma Cells Though Activation of the Sensors ATF6 and IRE1

Common upstream signaling pathways are involved in ER stress-induced apoptosis and autophagy. We thus tested the activation of three key molecules of these pathways: ATF6, PERK, and IRE1.

During ER stress, the inactive 90-kDa ATF6 protein undergoes proteolysis, leading to the release of a 50-kDa protein with transcriptional activity [25]. Treatment of BL cells with VT-1 resulted in the progressive cleavage of the inactive 90-kDa ATF6 into the active 50-kDa form (cleaved ATF6), with the complete disappearance of the full-length form after 8 h (Figure 2a). However, the kinetic of ATF6 activation was faster in Ramos than BL2 cells.

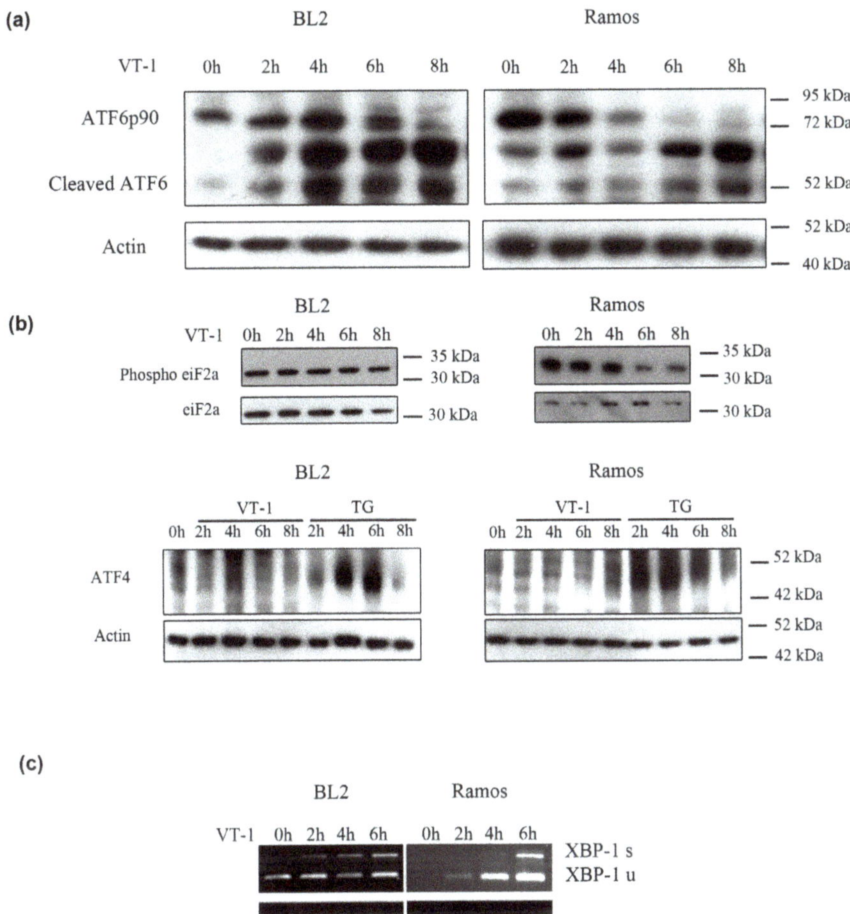

Figure 2. VT-1 activated ER stress sensors in BL cells. BL2 and Ramos cells were stimulated with VT-1 for 0, 2, 4, 6, or 8 h. (**a**) After treatment, cell lysates were prepared and ATF6 cleavage analyzed by western blotting. (**b**) Phospho-eIF2α, eIF2α, and ATF4 expression were also analyzed by western blotting. (**c**) After treatment, total RNA was isolated and amplified using XBP-1 specific primers. Amplification products were then incubated with the enzyme ApaLI. Unspliced XBP-1 (XBP-1 u cleaved by ApaLI into stackable ~280 and ~340 bp fragments) and spliced XBP-1 (XBP-1 s, 590 Bp, insensitive to ApaLI) were detected on agarose gels.

Activated PERK can phosphorylate the α subunits of eukaryotic initiation factor 2 (eIF2), a major regulator of mRNA translation [26]. We, therefore, measured the presence of phosphorylated eIF2α before and after VT-1 treatment by western-blot analysis (Figure 2b). There were no changes in eIF2α phosphorylation levels in BL2 cells, even after 8 h of treatment. On the contrary and intriguingly, the levels of phospho-eiF2α in Ramos cells decreased, beginning after 6 h of VT-1 treatment. The activation of PERK signaling also results in the attenuation of overall protein translation, concomitant with the induction of translation of only selective mRNA molecules, including those of activating transcription factor 4 (ATF4). We therefore also analyzed ATF4 expression after treatment of our BL cell lines with VT-1 (Figure 2b). We did not observe any increases in ATF4 levels, neither in BL2 nor Ramos cells, whereas ATF4 was clearly induced in these cells when they were treated with Thapsigargin, a well-known stress inducer. These results show that the PERK/eIF2α/ATF4 signaling pathway was not activated by VT-1 treatment of BL cells.

We then investigated activation of the IRE1 signaling pathway. IRE1 activation triggers splicing of the RNA transcript encoding the transcription factor XBP-1 [27]. One consequence of such splicing is the removal of an ApaLI site in the XBP-1 transcript. Thus, when incubated with ApaLI, the spliced XBP-1 transcript (XBP-1s) remains ~590 bp, whereas the unspliced XBP-1 transcript (XBP-1 u) is cleaved into ~280 and ~340-bp fragments (which are indistinguishable on gels). We thus extracted total mRNA of BL cells treated or not with VT-1 and amplified it using XBP-1 specific primers. The amplification products were then incubated with the enzyme ApaLI. XBP-1s was not present in untreated cells but gradually increased after treatment of both BL cell lines with VT-1, indicating IRE1 activation (Figure 2c). Contrary to ATF6, IRE1 appeared to be activated more quickly in BL2 than Ramos cells.

One of the hallmarks of ER stress-sensor activation is increased expression of the transcription factor CHOP. Thus, we analyzed CHOP expression at various times after VT-1 treatment both by qRT-PCR and western blotting. CHOP was constitutively expressed at low levels in non-treated cells and gradually increased after VT-1 treatment, both at the RNA (Figure 3a) and protein levels (Figure 3b).

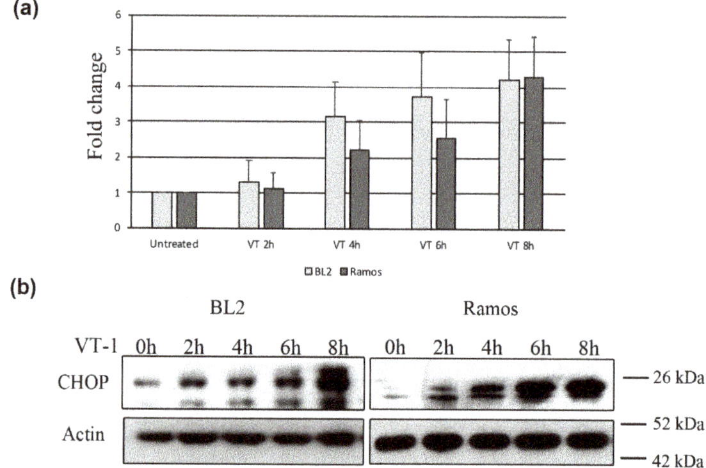

Figure 3. C/REB homologous protein (CHOP) expression increases in BL cells treated with VT-1. BL2 and Ramos cells were treated with VT-1 for 0, 2, 4, 6, or 8 h. After treatment, total RNA or protein extracts were prepared and CHOP induction analyzed by qRT-PCR (**a**) or western blotting (**b**).

Overall, our results show that VT-1 induces an ER stress response in both BL cell lines but only at the transcriptional regulation level (through ATF6 and IRE1) and not at the translational level (PERK/eIF2α/ATF4).

2.3. Silencing CHOP Protects BL2 but Not Ramos Cells from VT-1 Induced Apoptosis

Given the results to this point, we tested the involvement of CHOP in VT-1 induced apoptosis by inhibiting its expression. A lentiviral vector-based shRNA system was used to stably repress CHOP expression in Ramos and BL2 cells. Cells obtained with two different shRNA constructs or a control shRNA were treated with VT-1 and the CHOP mRNA level determined by qRT-PCR to verify the knockdown of CHOP. The induction of CHOP was clearly observed after VT-1 treatment of shCTRL cells but not BL2 shCHOP or Ramos shCHOP cells (Figure 4a). We then assessed apoptosis by flow cytometry. CHOP knockdown did not reduce VT-1-induced apoptosis in Ramos cells (53.6 ± 8%,

40.7 ± 5%, and 52.5 ± 4% apoptosis after 6 h of VT-1 treatment of Ramos shCTRL, Ramos shCHOP1, and Ramos shCHOP2, respectively) but protected BL2 cells, since the percentage of apoptosis was reduced by almost two times (59 ± 7%, 34.6 ± 4%, and 32.5 ±3% apoptosis after 6 h of VT-1 treatment of BL2 shCTRL, BL2 shCHOP1, and BL2 shCHOP2, respectively, Figure 4b).

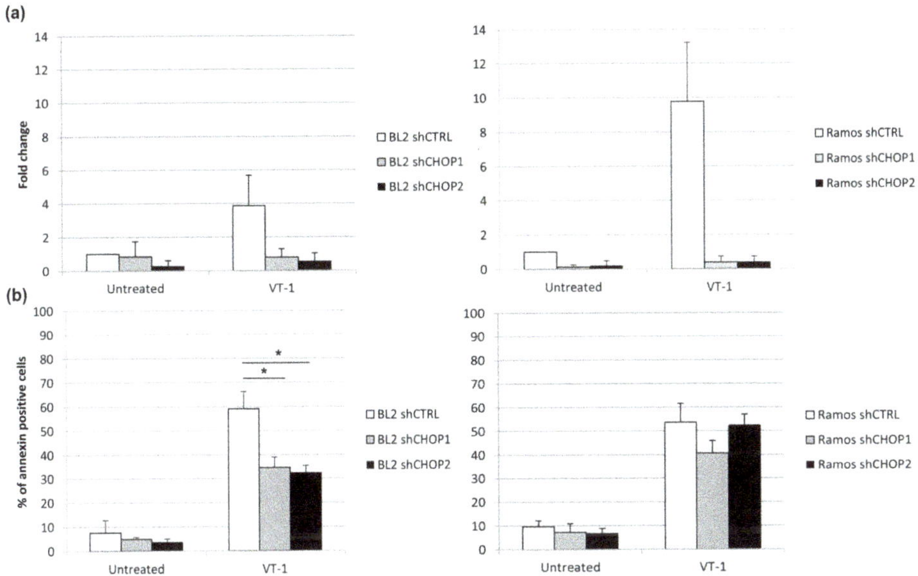

Figure 4. CHOP knockdown reduces VT-1-induced apoptosis in BL2 but not Ramos cells. After incubation of the cells with VT-1 for 6 h, the induction of CHOP was quantified by qRT-PCR (**a**) and the percentage of apoptotic cells analyzed by flow cytometry using annexinV/propidium iodide staining (**b**). Data are presented as the means ±SD from at least three independent experiments. Statistical test: Mann Whitney, * $p < 0.05$.

Previous reports have shown CHOP to be involved in ER stress-induced apoptosis through its ability to induce the upregulation of DR5 (death Receptor 5), which in turn activates caspase 8 and BAX [21]. Furthermore, others have shown that the upregulation of BIM induced by CHOP plays a central role in ER stress-triggered apoptosis, as well as downregulation of BCL-2 [28–30]. We therefore tested the expression of BCL-2 and BIM in the Ramos and BL2 cells before and after treatment with VT-1. BCL-2 was expressed in BL2 cells, but its level was not downregulated by VT-1 treatment (even slightly increased), whereas this protein was not present in Ramos cells (Figure 5). By contrast, BIM was not expressed in BL2 cells but was present in Ramos cells. However, VT-1 treatment clearly induced a decrease in the level of BIM in these cells. These results show that the role of CHOP in VT-1 induced apoptosis is not through BCL-2 or BIM regulation.

We then assessed DR5 expression in the BL2 and Ramos cell lines after VT-1 treatment. DR5 mRNA levels increased in the two cell lines after VT-1 treatment (Figure 6a). However, this result was not confirmed by the analysis of DR5 expression at the cell surface by flow cytometry. Indeed, up to 100% of the cells clearly expressed DR5 before treatment and its expression gradually decreased between 4 and 8 h of VT-1 treatment, as shown by the mean fluorescence intensity (MFI) reported under each graph in Figure 6b (MFI decrease from 371 to 324 for BL2 cells and from 336 to 295 for Ramos cells). Overall, our results suggest that in certain BL cells, CHOP participates in VT-1-induced apoptotic signaling but not through already-known mechanisms involving the BCL-2 family or DR5.

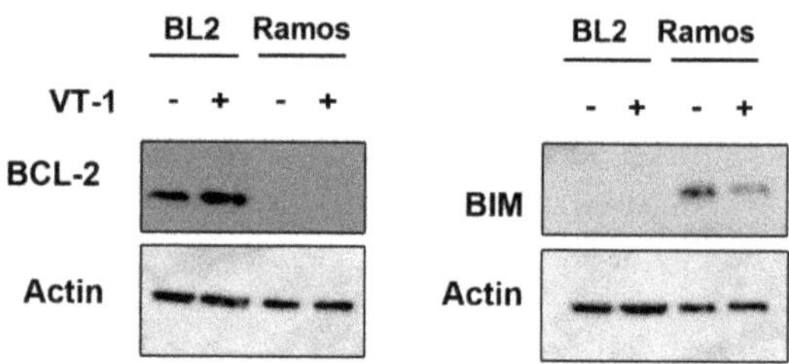

Figure 5. BCL-2 and BIM expression in BL cell lines by western blotting. Cells were incubated or not with VT-1 for 6 h.

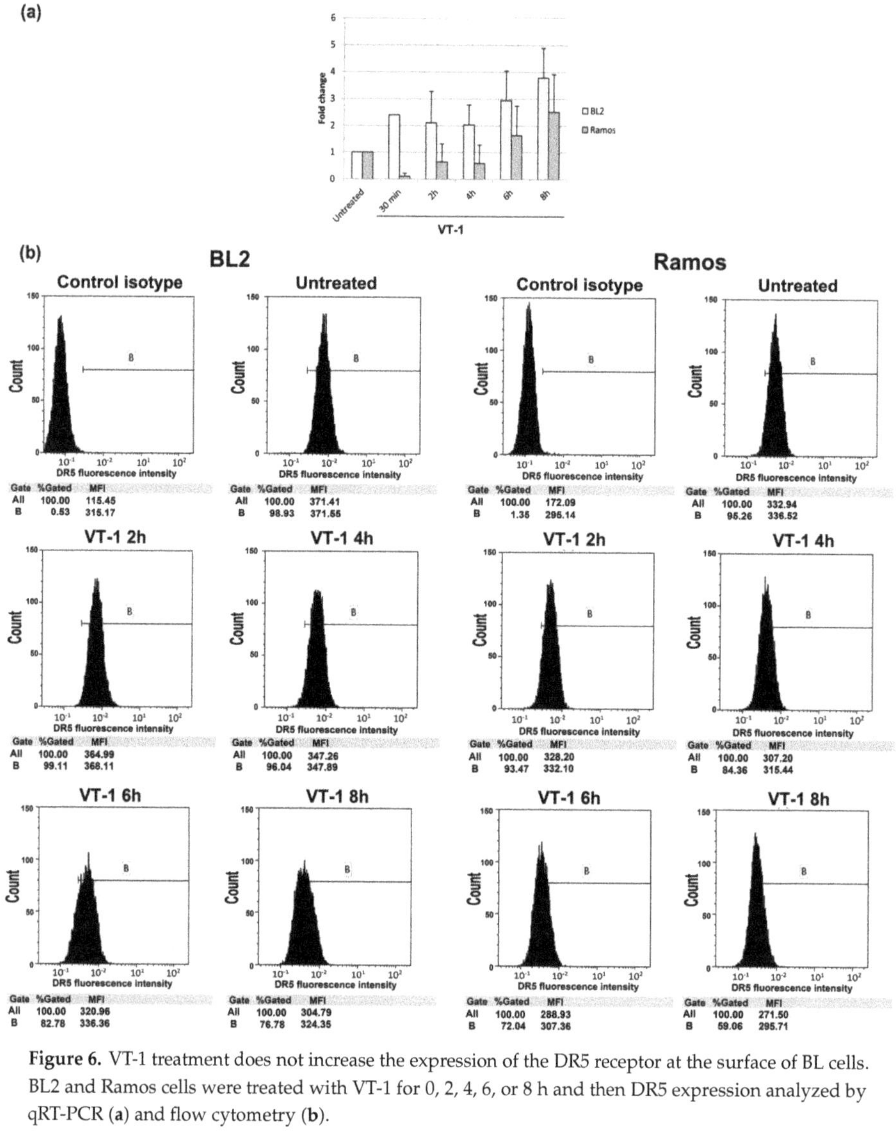

(a)

(b)

Figure 6. VT-1 treatment does not increase the expression of the DR5 receptor at the surface of BL cells. BL2 and Ramos cells were treated with VT-1 for 0, 2, 4, 6, or 8 h and then DR5 expression analyzed by qRT-PCR (**a**) and flow cytometry (**b**).

2.4. Calpain Activation Is Not Involved in VT-1-Induced Apoptosis

We further investigated the role of ER stress in VT-1-induced apoptosis by assessing the role of calcium in this pathway. Indeed, the activation of ER stress is frequently accompanied by calcium release into the cytosol and an increase in the calcium concentration has also been implicated in the induction of apoptosis [31,32]. Notably, Calpain, a cysteine protease activated by elevated intracellular Ca^{2+} levels induced by ER stress can act as an alternative pathway to activate caspases. We determined whether calpain activation is involved in VT-1-induced apoptosis by pretreating cells with ALLM (calpain and cathepsin inhibitor) for 30 min and then induced apoptosis with VT-1. After 6 h of treatment, the cells were analyzed by flow cytometry after annexin V/PI labeling. ALLM had no effect on VT-1-induced apoptosis either in BL2 nor in Ramos cells (Figure 7). These results suggest that the induction of apoptosis by VT-1 in BL cells relies on a Ca^{2+}-independent signaling pathway.

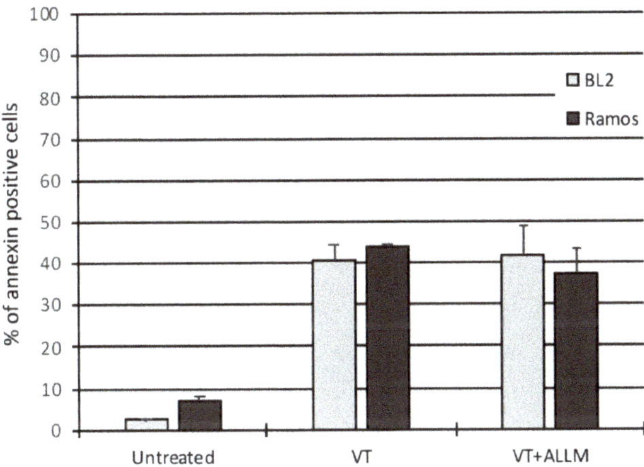

Figure 7. VT-1-induced apoptosis is independent of calpain activation. Cells were pretreated with ALLM (calpain and cathepsin inhibitor) for 30 min and then VT-1. After 6 h of treatment, cells were analyzed by flow cytometry using annexin V/PI.

3. Discussion

Numerous reports have shown that VT-1 is able to trigger multiple effector pathways that, for the vast majority, lead to apoptosis of the cells [13–15]. Here, we found that treating BL cells with VT-1 induced the activation of IRE1 and ATF6, the two main sensors of ER stress that operate at the transcriptional level, but not the third one, PERK, which is directly involved in globally shutting off mRNA translation. We also showed that VT-1 treatment results in the upregulation of CHOP, which is normally the point at which ER stress pathways switch from the restoration of homeostasis to programmed cell death. However, we observed differential roles for CHOP, which appears to be part of the VT-1-induced apoptotic pathway in BL2 cells, whereas it is not implicated in the death process of Ramos cells. On the contrary, a selective autophagy pathway called ER-phagy restrained VT-1-induced apoptosis in these cells (Figure 8). It is possible that ER-phagy plays a protective role by preventing the toxic accumulation of unfolded proteins in the ER resulting from the VT-1-induced inhibition of protein synthesis or by altering intracellular toxin routing (and promoting its proteolytic degradation).

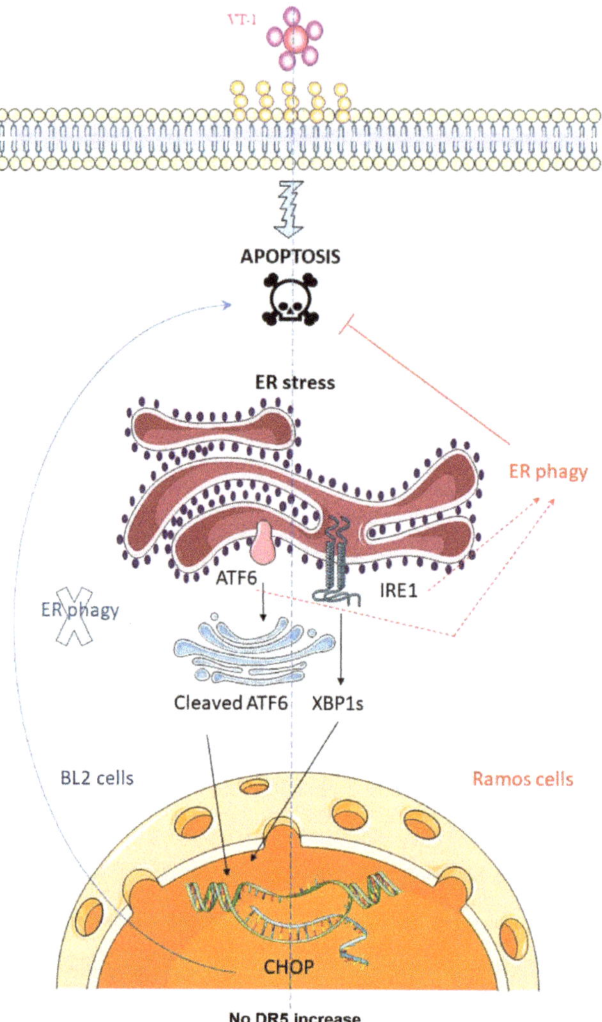

Figure 8. VT-1-induced ER stress triggers apoptotic or survival pathways in Burkitt lymphoma cells. We showed previously that the signal transduction pathway induced by VT-1 in BL cells is a relatively conventional caspase- and mitochondria-dependent pathway [17–19]. Here, we report that treating cells with VT-1 induces the ER stress response by activation of the ER stress sensors IRE1 and ATF6, followed by increased expression of the transcription factor CHOP. Interestingly, we observed differential roles for CHOP, which appears to be part of the VT-1-induced apoptotic pathway in BL2 cells, whereas it is not implicated in the death process of Ramos cells. On the contrary, ER-phagy which occurred in these cells restrained VT-1-induced apoptosis. The role of CHOP in Ramos cells and the pathway (ATF6 or/and IRE1) involved in triggering ER-phagy are still not known. Dashed lines indicate possible mechanisms, solid lines indicate demonstrated effects.

UPR, the cellular response to ER stress, is intrinsically related to autophagy, which acts as a cytoprotective factor and proceeds through two interconnected pathways, ER stress-mediated autophagy and ER-phagy. ER-phagy is a recently identified form of selective autophagy and there are still many questions about its molecular mechanisms and physiological role. However, it is generally accepted that the core autophagy machinery is required for its activation [33]. Others have studied the role of autophagy in Stxs-induced apoptosis [34–36] and all have shown that autophagy inhibitors protect against toxin cytotoxicity, suggesting that autophagy participates in the cell death mechanism. Interestingly, Lee et al. showed that autophagy is induced by Stxs, both in toxin-sensitive and toxin-resistant cells, but that calpains and caspases can cleave ATG5 and BECLIN only in sensitive cells and thus transform a pro-survival autophagic response to an apoptotic response. The discrepancies between these results and ours concerning the role of autophagy in VT/Stx-induced apoptosis may be due to the different cell types studied.

In neuroblastoma cells, Ogata et al. showed that the ER stressors thapsigargin and tunicamycin induce the formation of autophagosomes via the IRE1/JNK pathway, whereas PERK and ATF6 appear not to be involved [37], and that such IRE1/JNK-induced autophagy protects the cells against death induced by ER stress. In this context, it is interesting to note that starvation-induced autophagy is mediated by the specific phosphorylation of ER-localized BCL-2 by JNK1, which leads to disruption of the BCL2/BECLIN complex [38]. On the other hand, others have shown that phosphorylation of BCL-2 can disrupt the association of BCL-2 either with BECLIN or BAX, thus contributing to both autophagy and apoptosis, and that these events can occur sequentially in the same cell [39]. Finally, it has also been shown that the IRE1/XBP-1s axis induces autophagy, with increased conversion of LC3I to LC3II and increased expression of BECLIN [40]. In our BL model, we show that VT-1 induces the ER stress response through the activation of ATF6 and IRE1/XBP-1s, suggesting that one of these two arms is responsible for the induction of ER-phagy. However, there was no activation or involvement of calpain, no increased expression or cleavage of BECLIN, and no change in BCL-2 phosphorylation (data not show) in our cell lines after VT-1 treatment, consistent with a pro-survival effect of autophagy but not through activation of the IRE1/JNK/XBP-1s pathways.

ATF6 has also been shown to contribute to ER expansion [41] and autophagy, notably through its interaction with C/EBP-β and the ability of this complex to induce the expression of DAPK1 [42], a kinase that can phosphorylate BECLIN, thus releasing it from BCL-XL and resulting in autophagosome formation [43]. In our model, we were unable to detect a change in BECLIN phosphorylation (data not shown) after VT-1 treatment, which does not favor a role for DAPK1. ATF6 may also be involved in the initiation of autophagy by upregulating the expression of BIP [44]. Indeed, BIP has been shown to modulate AKT signaling, which is a well-known regulator of mTOR-mediated autophagy [45]. Since the IRE1 arm does not appear to be involved in the induction of ER-phagy in our BL model, it would certainly be worthwhile to investigate the role of ATF6 in this process.

We show that CHOP is upregulated in BL2 cells during VT-1 treatment and that it participates in the induction of the apoptotic pathway. However, we also show that the effect of CHOP on apoptosis is not due—contrary to what has been previously reported [21]—to the downregulation of BCL-2 and upregulation of DR5. This raises the question of whether CHOP is directly involved in the apoptotic pathways or whether it inhibits the protection conferred by ER-phagy. In our case, it is possible that CHOP induced a pathway that inhibited ER-phagy in BL2 cells but that this pathway was not activated in Ramos cells. Others have suggested that CHOP may promote ER stress-induced apoptosis via the inhibition of autophagy. They showed that specific shRNA inhibition of CHOP in hepatocellular carcinoma resulted in enhanced tunicamycin-induced autophagy (shown by increased LC3 II expression) and reduced apoptosis [46]. On the other hand, it has been shown that CHOP can directly promote PUMA (p53 upregulated modulator of apoptosis) expression in certain circumstances of induced ER stress and that the CHOP/PUMA axis can synergize with the classical apoptotic process [47]. The role of CHOP in VT-1-induced apoptosis of BL cells is yet to be elucidated.

We previously reported that treatment of BL cell lines with VT-1 induces a caspase- and mitochondria-dependent apoptotic pathway in which BID is essential [17–19]. Here, VT-1 also simultaneously induced ER stress, which activated different signaling pathways, depending on the cell line. The kinetics of ER stress induction (between 2 and 4 h) after VT-1 treatment would be compatible with an effect on caspase 8 activation, as previously reported in other models [48]. However, we previously showed that caspase 8 is most likely controlled by c-FLIP$_L$ degradation, which occurs via the ubiquitin–proteasome pathway [17]. Furthermore, the fact that ER stress contributed to either survival or death in different BL cell lines suggests that it is not necessary for VT-1 induced apoptosis.

Because Gb3 functions as a receptor for Stxs and its expression is deregulated in various malignancies [8–10], several engineered Stxs are currently under investigation as potential anti-cancer agents. However, the utility of Stxs is considered to be limited, as it induces endothelial cell damage and is responsible for HUS observed in patients infected with Shiga toxin-producing *Escherichia coli* [12]. Furthermore, another detrimental effect is that Stxs can damage human hematopoietic progenitor cells since they expressed Gb3 and Gb4 receptors at the primary stage of erythropoietic differentiation [49]. A better understanding of the molecular mechanisms that determine the outcome of Stxs-induced ER stress and autophagy in each cell type will certainly offer new opportunities to improve their potential as cancer therapies. Indeed, it is now clear that, depending on the cell type, Stxs can induce different signaling pathways, such as autophagy and ER stress, which activate both pro-survival and pro-death mechanism and thereby play a dual role in VT-1-mediated killing. It is thus possible that stimulating or blocking certain pathways could improve the specific action of the toxin, depending on the cell type (normal versus tumor). Future investigations will be necessary to more precisely determine whether combining VT-1 with molecules that enhance autophagy could protect normal cells while killing tumor cells.

4. Materials and Methods

4.1. Cell Lines

The Ramos cell line was obtained from the American type culture collection (ATCC-CRL-1596, Rockville, MD, USA). The BL2 cell line was originally established from a case of BL and kindly provided by the International Agency for Research on Cancer (IARC, Lyon, France). Both cell lines were previously tested for Gb3 expression and VT-1 sensitivity [15,17–19]. Mission®shRNA lentiviral transduction particles (Sigma Aldrich, Saint-Quentin Fallavier, France) were used to suppress CHOP gene expression (SHCLNV-TRCN00000007263, 7264) and Mission®non-target shRNA control lentiviral transduction particles (Sigma Aldrich, Saint-Quentin Fallavier, France) were used as a control (SHC002V). The 2×10^6 BL cells were transduced with lentiviral particles (multiplicity of infection (MOI) of 15) in fresh media distributed in 6 wells plates. Cells were then incubated for 24 h at 37 °C in a humidified incubator in an atmosphere of 5% CO_2. The media containing lentiviral particles was then replaced by fresh media and incubation was continued for 24 h. Cells were then cultured in media supplemented with selection agent (0.6 µg/mL puromycin) for several weeks. Two CHOP-repressed cell lines were established for each parental cell line: Ramos shCHOP1, Ramos shCHOP2 and BL2 shCHOP1, BL2 shCHOP2. Two control cell lines: Ramos shCTRL and BL2 shCTRL (which behave like the parental Ramos and BL2 cell lines) were also established.

All cell lines were cultured in RPMI 1640 medium (Life technology, Villebon sur Yvette, France) containing 2 mM l-glutamine, 1 mM sodium pyruvate, 20 mM glucose, 100 U/mL penicillin, and 100 µg/mL streptomycin and supplemented with 7% heat-inactivated fetal calf serum.

4.2. Total Extract Preparation

Aliquots of 1×10^6 cells were pelleted, solubilized in ice-cold RIPA buffer (150 mM NaCl, 50 mM Tris, pH 7.5, 5 mM EDTA, 0.5% NP40, 0.5% NaDOC, 0.1% SDS, complete protease inhibitor) and sonicated. Then the samples were analyzed by western blotting.

4.3. Western-Blot Analysis

The protocol for western-blotting was described elsewhere [17]. Antibodies used: ATF6 (IMGENEX, clone 70B1413.1, Noyal Chatillon sur Seiche, France), Phospho- eIF2α (Abcam, clone E90,

Cambridge, UK), eIF2α (Santa Cruz, clone D-3, Heidelberg, Germany), CHOP (Santa Cruz, clone B-3), and ATF4 (Santa Cruz, clone c-20).

4.4. Detection of XBP-1 mRNA Splicing

After total RNA extraction, RT-PCR products of XBP-1 mRNA were obtained using primers 5′-cggtgcgcggtgcgtagtctgga-3′ (sense) and 5′-tgaggggctgagaggtgcttcct-3′ (anti-sense). Upon activation of XBP-1 mRNA, a 26 bp fragment containing an Apa-LI site is spliced. Therefore, the RT-PCR products were digested with Apa-LI to distinguish the spliced form from the unspliced form. The inactive form was revealed by electrophoresis as two cleaved fragments and the active form as a non-cleaved fragment.

4.5. Analysis of DR5 Cell-Surface Expression

BL cells were incubated with or without VT-1 (5 ng/mL) for 0, 2, 4, 6, or 8 h. Cells were then incubated with antibodies directed against membrane-bound DR5 (Santa Cruz, clone DJR2-4) for at least 30 min at 4 °C. Following incubation, cells were washed and then incubated with secondary antibody (goat anti-mouse Alexa 488, Life technologies) for at least 30 min at 4 °C. Finally, cells were washed and suspended in 0.2 mL PBS. Cell fluorescence was measured by flow cytometry (Accuri C6 cytometer, Becton-Dickinson, Pont-de-Claix, France). Untreated cells and cells incubated with the isotype control and then the fluorescein-labeled secondary antibody served as controls.

4.6. Cell Death Measurement

Cell death was assessed by labeling cells with annexin V-FITC and propidium iodide (PI). Aliquots of 1×10^6 cells were incubated for 6 h at 37 °C with VT-1 (5 ng/mL) in 1 mL of complete RPMI medium. The staining protocol was described previously [17]

4.7. Electron Microscopy

Cells were fixed in 2% glutaraldehyde in 0.1 M cacodylate buffer, pH 7.4 for 1 h at room temperature. They were then post-fixed for 1 h in 1% osmium tetroxide mixed with 1.5% potassium ferrocyanide in the same buffer. The samples were then exposed to successive baths of increasing ethanol concentration for dehydration and embedded in epoxy resin (Embed 812, EMS 14120). Once the resin had cured, ultra-thin sections (70 nm) were performed and collected on copper grids coated with collodion-carbon (EMS, G200-Cu). They were stained with a 2% aqueous solution of uranyl acetate, followed by a staining with lead citrate in Reynold's solution. Finally, the sections were observed with the Zeiss 902 electron microscope (Carl Zeiss microscopy GmbH, Jena, Germany), using the Megaview III CCD camera (Olympus Soft Imaging Solutions, GmbH, Münster, Germany).

4.8. Data Analysis and Statistics

All values for statistical significance represent the mean ± standard deviation (s.d.). Statistical analyses were performed using the non-parametric Mann Whitney test for side-by-side comparisons. Differences were considered to be statistically significance for $p < 0.05$.

Author Contributions: Conceptualization, A.R.; methodology, A.R.; validation, J.D., C.P.-D., E.L.C. and A.R.; formal analysis, A.R.; investigation, A.R., C.P.-D., and J.D.; writing—original draft preparation, A.R.; writing—review and editing, J.W.; visualization, A.R. and C.P.-D.; supervision, A.R. and J.W.; project administration, A.R. and J.W.; funding acquisition, J.W. All authors have read and agreed to the published version of the manuscript.

Funding: This research was funded by the Fondation de France (#: 2014 00047509), the FCS campus Paris Saclay (AAP prematuration 2015-16) and the Ligue Contre le Cancer (doctoral fellowship to JD).

Acknowledgments: The authors thank Yann Lecluse (Imaging and Cytometry Platform, Gustave Roussy) for expert technical assistance in performing the flow cytometry analyses.

Conflicts of Interest: The authors declare no conflict of interest.

References

1. Boyer, O.; Niaudet, P. Hemolytic uremic syndrome: New developments in pathogenesis and treatment. *Int. J. Nephrol.* **2011**, *2011*, 908407. [CrossRef]

2. Joseph, A.; Cointe, A.; Mariani Kurkdjian, P.; Rafat, C.; Hertig, A. Shiga Toxin-Associated Hemolytic Uremic Syndrome: A Narrative Review. *Toxins* **2020**, *12*, 67. [CrossRef]

3. Jacewicz, M.; Clausen, H.; Nudelman, E.; Donohue-Rolfe, A.; Keusch, G.T. Pathogenesis of shigella diarrhea. XI. Isolation of a shigella toxin-binding glycolipid from rabbit jejunum and HeLa cells and its identification as globotriaosylceramide. *J. Exp. Med.* **1986**, *163*, 1391–1404. [CrossRef] [PubMed]

4. DeGrandis, S.; Law, H.; Brunton, J.; Gyles, C.; Lingwood, C.A. Globotetraosylceramide is recognized by the pig edema disease toxin. *J. Biol. Chem.* **1989**, *264*, 12520–12525. [PubMed]

5. Johannes, L. Shiga Toxin-A Model for Glycolipid-Dependent and Lectin-Driven Endocytosis. *Toxins* **2017**, *9*, 340. [CrossRef] [PubMed]

6. Bergan, J.; Dyve Lingelem, A.B.; Simm, R.; Skotland, T.; Sandvig, K. Shiga toxins. *Toxicon* **2012**, *60*, 1085–1107. [CrossRef]

7. Nudelman, E.; Kannagi, R.; Hakomori, S.; Parsons, M.; Lipinski, M.; Wiels, J.; Fellous, M.; Tursz, T. A glycolipid antigen associated with Burkitt lymphoma defined by a monoclonal antibody. *Science* **1983**, *220*, 509–511. [CrossRef]

8. Arab, S.; Russel, E.; Chapman, W.B.; Rosen, B.; Lingwood, C.A. Expression of the verotoxin receptor glycolipid, globotriaosylceramide, in ovarian hyperplasias. *Oncol Res.* **1997**, *9*, 553–563.

9. Gupta, V.; Bhinge, K.N.; Hosain, S.B.; Xiong, K.; Gu, X.; Shi, R.; Ho, M.Y.; Khoo, K.H.; Li, S.C.; Li, Y.T.; et al. Ceramide glycosylation by glucosylceramide synthase selectively maintains the properties of breast cancer stem cells. *J. Biol. Chem.* **2012**, *287*, 37195–37205. [CrossRef]

10. Ohyama, C.; Fukushi, Y.; Satoh, M.; Saitoh, S.; Orikasa, S.; Nudelman, E.; Straud, M.; Hakomori, S. Changes in glycolipid expression in human testicular tumor. *Int. J. Cancer* **1990**, *45*, 1040–1044. [CrossRef]

11. Bien, T.; Perl, M.; Machmüller, A.C.; Nitsche, U.; Conrad, A.; Johannes, L.; Müthing, J.; Soltwisch, J.; Janßen, K.P.; Dreisewerd, K. MALDI-2 Mass Spectrometry and Immunohistochemistry Imaging of Gb3Cer, Gb4Cer, and Further Glycosphingolipids in Human Colorectal Cancer Tissue. *Anal. Chem.* **2020**. [CrossRef] [PubMed]

12. Engedal, N.; Skotland, T.; Torgersen, M.L.; Sandvig, K. Shiga toxin and its use in targeted cancer therapy and imaging. *Microb. Biotechnol.* **2011**, *4*, 32–46. [CrossRef]

13. Lee, S.Y.; Cherla, R.P.; Caliskan, I.; Tesh, V.L. Shiga toxin 1 induces apoptosis in the human myelogenous leukemia cell line THP-1 by a caspase-8-dependent, tumor necrosis factor receptor-independent mechanism. *Infect. Immun.* **2005**, *73*, 5115–5126. [CrossRef] [PubMed]

14. Ching, J.C.; Jones, N.L.; Ceponis, P.J.; Karmali, M.A.; Sherman, P.M. Escherichia coli shiga-like toxins induce apoptosis and cleavage of poly(ADP-ribose) polymerase via in vitro activation of caspases. *Infect. Immun.* **2002**, *70*, 4669–4677. [CrossRef] [PubMed]

15. Mangeney, M.; Lingwood, C.A.; Taga, S.; Caillou, B.; Tursz, T.; Wiels, J. Apoptosis induced in Burkitt's lymphoma cells via Gb3/CD77, a glycolipid antigen. *Cancer Res.* **1993**, *53*, 5314–5319. [PubMed]

16. Fujii, J.; Matsui, T.; Heatherly, D.P.; Schlegel, K.H.; Lobo, P.I.; Yutsudo, T.; Ciraolo, G.M.; Morris, R.E.; Obrig, T. Rapid apoptosis induced by Shiga toxin in HeLa cells. *Infect. Immun.* **2003**, *71*, 2724–2735. [CrossRef] [PubMed]

17. Garibal, J.; Hollville, E.; Renouf, B.; Tetaud, C.; Wiels, J. Caspase-8-mediated cleavage of Bid and protein phosphatase 2A-mediated activation of Bax are necessary for Verotoxin-1-induced apoptosis in Burkitt's lymphoma cells. *Cell Signal.* **2010**, *22*, 467–475. [CrossRef]

18. Tetaud, C.; Falguieres, T.; Carlier, K.; Lecluse, Y.; Garibal, J.; Coulaud, D.; Busson, P.; Steffensen, R.; Clausen, H.; Johannes, L.; et al. Two distinct Gb3/CD77 signaling pathways leading to apoptosis are triggered by anti-Gb3/CD77 mAb and verotoxin-1. *J. Biol. Chem.* **2003**, *278*, 45200–45208. [CrossRef]

19. Debernardi, J.; Hollville, E.; Lipinski, M.; Wiels, J.; Robert, A. Differential role of FL-BID and t-BID during verotoxin-1-induced apoptosis in Burkitt's lymphoma cells. *Oncogene* **2018**, *37*, 2410–2421. [CrossRef]

20. Bertolotti, A.; Zhang, Y.; Hendershot, L.M.; Harding, H.P.; Ron, D. Dynamic interaction of BiP and ER stress transducers in the unfolded-protein response. *Nat. Cell Biol.* **2000**, *2*, 326–332. [CrossRef]

21. Lee, S.Y.; Lee, M.S.; Cherla, R.P.; Tesh, V.L. Shiga toxin 1 induces apoptosis through the endoplasmic reticulum stress response in human monocytic cells. *Cell Microbiol.* **2008**, *10*, 770–780. [CrossRef]

22. Lee, M.S.; Cherla, R.P.; Lentz, E.K.; Leyva-Illades, D.; Tesh, V.L. Signaling through C/EBP homologous protein and death receptor 5 and calpain activation differentially regulate THP-1 cell maturation-dependent apoptosis induced by Shiga toxin type 1. *Infect. Immun.* **2010**, *78*, 3378–3391. [CrossRef]

23. Fujii, J.; Wood, K.; Matsuda, F.; Carneiro-Filho, B.A.; Schlegel, K.H.; Yutsudo, T.; Binnington-Boyd, B.; Lingwood, C.A.; Obata, F.; Kim, K.S.; et al. Shiga toxin 2 causes apoptosis in human brain microvascular endothelial cells via C/EBP homologous protein. *Infect. Immun.* **2008**, *76*, 3679–3689. [CrossRef]

24. Hetz, C. The unfolded protein response: controlling cell fate decisions under ER stress and beyond. *Nat. Rev. Mol. Cell Biol.* **2012**, *13*, 89–102. [CrossRef]

25. Haze, K.; Yoshida, H.; Yanagi, H.; Yura, T.; Mori, K. Mammalian transcription factor ATF6 is synthesized as a transmembrane protein and activated by proteolysis in response to endoplasmic reticulum stress. *Mol. Biol. Cell* **1999**, *10*, 3787–3799. [CrossRef]

26. Harding, H.P.; Zhang, Y.; Bertolotti, A.; Zeng, H.; Ron, D. Perk is essential for translational regulation and cell survival during the unfolded protein response. *Mol. Cell* **2000**, *5*, 897–904. [CrossRef]

27. Yoshida, H.; Matsui, T.; Yamamoto, A.; Okada, T.; Mori, K. XBP1 mRNA is induced by ATF6 and spliced by IRE1 in response to ER stress to produce a highly active transcription factor. *Cell* **2001**, *107*, 881–891. [CrossRef]

28. McCullough, K.D.; Martindale, J.L.; Klotz, L.O.; Aw, T.Y.; Holbrook, N.J. Gadd153 sensitizes cells to endoplasmic reticulum stress by down-regulating Bcl2 and perturbing the cellular redox state. *Mol. Cell Biol.* **2001**, *21*, 1249–1259. [CrossRef]

29. Puthalakath, H.; O'Reilly, L.A.; Gunn, P.; Lee, L.; Kelly, P.N.; Huntington, N.D.; Hughes, P.D.; Michalak, E.M.; McKimm-Breschkin, J.; Motoyama, N.; et al. ER stress triggers apoptosis by activating BH3-only protein Bim. *Cell* **2007**, *129*, 1337–1349. [CrossRef]

30. Yamaguchi, H.; Wang, H.G. CHOP is involved in endoplasmic reticulum stress-induced apoptosis by enhancing DR5 expression in human carcinoma cells. *J. Biol. Chem.* **2004**, *279*, 45495–45502. [CrossRef]

31. Carreras-Sureda, A.; Pihán, P.; Hetz, C. Calcium signaling at the endoplasmic reticulum: fine-tuning stress responses. *Cell Calcium* **2018**, *70*, 24–31. [CrossRef]

32. Orrenius, S.; Zhivotovsky, B.; Nicotera, P. Regulation of cell death: the calcium-apoptosis link. *Nat. Rev. Mol. Cell Biol.* **2003**, *4*, 552–565. [CrossRef]

33. Song, S.; Tan, J.; Miao, Y.; Zhang, Q. Crosstalk of ER stress-mediated autophagy and ER-phagy: Involvement of UPR and the core autophagy machinery. *J. Cell Physiol.* **2018**, *233*, 3867–3874. [CrossRef]

34. Lee, M.S.; Cherla, R.P.; Jenson, M.H.; Leyva-Illades, D.; Martinez-Moczygemba, M.; Tesh, V.L. Shiga toxins induce autophagy leading to differential signalling pathways in toxin-sensitive and toxin-resistant human cells. *Cell Microbiol.* **2011**, *13*, 1479–1496. [CrossRef]

35. Sandvig, K.; Garred, O.; Prydz, K.; Kozlov, J.V.; Hansen, S.H.; van Deurs, B. Retrograde transport of endocytosed Shiga toxin to the endoplasmic reticulum. *Nature* **1992**, *358*, 510–512. [CrossRef]

36. Tang, B.; Li, Q.; Zhao, X.H.; Wang, H.G.; Li, N.; Fang, Y.; Wang, K.; Jia, Y.P.; Zhu, P.; Gu, J.; et al. Shiga toxins induce autophagic cell death in intestinal epithelial cells via the endoplasmic reticulum stress pathway. *Autophagy* **2015**, *11*, 344–354. [CrossRef]

37. Ogata, M.; Hino, S.; Saito, A.; Morikawa, K.; Kondo, S.; Kanemoto, S.; Murakami, T.; Taniguchi, M.; Tanii, I.; Yoshinaga, K.; et al. Autophagy is activated for cell survival after endoplasmic reticulum stress. *Mol. Cell Biol.* **2006**, *26*, 9220–9231. [CrossRef]

38. Wei, Y.; Pattingre, S.; Sinha, S.; Bassik, M.; Levine, B. JNK1-mediated phosphorylation of Bcl-2 regulates starvation-induced autophagy. *Mol. Cell* **2008**, *30*, 678–688. [CrossRef]

39. Wei, Y.; Sinha, S.; Levine, B. Dual role of JNK1-mediated phosphorylation of Bcl-2 in autophagy and apoptosis regulation. *Autophagy* **2008**, *4*, 949–951. [CrossRef]

40. Margariti, A.; Li, H.; Chen, T.; Martin, D.; Vizcay-Barrena, G.; Alam, S.; Karamariti, E.; Xiao, Q.; Zampetaki, A.; Zhang, Z.; et al. XBP1 mRNA splicing triggers an autophagic response in endothelial cells through BECLIN-1 transcriptional activation. *J. Biol. Chem.* **2013**, *288*, 859–872. [CrossRef]

41. Bommiasamy, H.; Back, S.H.; Fagone, P.; Lee, K.; Meshinchi, S.; Vink, E.; Sriburi, R.; Frank, M.; Jackowski, S.; Kaufman, R.J.; et al. ATF6alpha induces XBP1-independent expansion of the endoplasmic reticulum. *J. Cell Sci.* **2009**, *122*, 1626–1636. [CrossRef] [PubMed]

42. Gade, P.; Ramachandran, G.; Maachani, U.B.; Rizzo, M.A.; Okada, T.; Prywes, R.; Cross, A.S.; Mori, K.; Kalvakolanu, D.V. An IFN-gamma-stimulated ATF6-C/EBP-beta-signaling pathway critical for the expression of Death Associated Protein Kinase 1 and induction of autophagy. *Proc. Natl. Acad. Sci. USA* **2012**, *109*, 10316–10321. [CrossRef] [PubMed]

43. Zalckvar, E.; Berissi, H.; Mizrachy, L.; Idelchuk, Y.; Koren, I.; Eisenstein, M.; Sabanay, H.; Pinkas-Kramarski, R.; Kimchi, A. DAP-kinase-mediated phosphorylation on the BH3 domain of beclin 1 promotes dissociation of beclin 1 from Bcl-XL and induction of autophagy. *EMBO Rep.* **2009**, *10*, 285–292. [CrossRef]

44. Yoshida, H.; Haze, K.; Yanagi, H.; Yura, T.; Mori, K. Identification of the cis-acting endoplasmic reticulum stress response element responsible for transcriptional induction of mammalian glucose-regulated proteins. Involvement of basic leucine zipper transcription factors. *J. Biol. Chem.* **1998**, *273*, 33741–33749. [CrossRef]

45. Yung, H.W.; Charnock-Jones, D.S.; Burton, G.J. Regulation of AKT phosphorylation at Ser473 and Thr308 by endoplasmic reticulum stress modulates substrate specificity in a severity dependent manner. *PLoS ONE* **2011**, *6*, e17894. [CrossRef]

46. Lei, Y.; Wang, S.; Ren, B.; Wang, J.; Chen, J.; Lu, J.; Zhan, S.; Fu, Y.; Huang, L.; Tan, J. CHOP favors endoplasmic reticulum stress-induced apoptosis in hepatocellular carcinoma cells via inhibition of autophagy. *PLoS ONE* **2017**, *12*, e0183680. [CrossRef]

47. Lee, Y.S.; Lee, D.H.; Choudry, H.A.; Bartlett, D.L.; Lee, Y.J. Ferroptosis-Induced Endoplasmic Reticulum Stress: Cross-talk between Ferroptosis and Apoptosis. *Mol. Cancer Res.* **2018**, *16*, 1073–1076. [CrossRef]

48. Iurlaro, R.; Muñoz-Pinedo, C. Cell death induced by endoplasmic reticulum stress. *FEBS J.* **2016**, *283*, 2640–2652. [CrossRef]

49. Betz, J.; Dorn, I.; Kouzel, I.U.; Bauwens, A.; Meisen, I.; Kemper, B.; Bielaszewska, M.; Mormann, M.; Weymann, L.; Sibrowski, W.; et al. Shiga toxin of enterohaemorrhagic Escherichia coli directly injures developing human erythrocytes. *Cell Microbiol.* **2016**, *18*, 1339–1348. [CrossRef]

 *toxins*

Article

Cnf1 Variants Endowed with the Ability to Cross the Blood–Brain Barrier: A New Potential Therapeutic Strategy for Glioblastoma

Andrea Colarusso [1], Zaira Maroccia [2], Ermenegilda Parrilli [1], Elena Angela Pia Germinario [2], Andrea Fortuna [2], Stefano Loizzo [2], Laura Ricceri [2], Maria Luisa Tutino [1], Carla Fiorentini [2,3] and Alessia Fabbri [2,*]

[1]   Department of Chemical Sciences, University of Naples Federico II, Complesso Universitario M. S. Angelo, Via Cintia, 80126 Napoli, Italy; andrea.colarusso@unina.it (A.C.); erparril@unina.it (E.P.); tutino@unina.it (M.L.T.)

[2]   Istituto Superiore di Sanità, Viale Regina Elena 299, 00161 Rome, Italy; zaira.maroccia@iss.it (Z.M.); elena.germinario@iss.it (E.A.P.G.); andrea.fortuna@iss.it (A.F.); stefano.loizzo@iss.it (S.L.); laura.ricceri@iss.it (L.R.); carla.fiorentini@iss.it (C.F.)

[3]   Association for Research on Integrative Oncological Therapies (ARTOI), 00165 Rome, Italy

*   Correspondence: alessia.fabbri@iss.it; Tel.: +39-06-4990-2939

Received: 14 April 2020; Accepted: 2 May 2020; Published: 4 May 2020

Abstract: Among gliomas, primary tumors originating from glial cells, glioblastoma (GBM) identified as WHO grade IV glioma, is the most common and aggressive malignant brain tumor. We have previously shown that the *Escherichia coli* protein toxin cytotoxic necrotizing factor 1 (CNF1) is remarkably effective as an anti-neoplastic agent in a mouse model of glioma, reducing the tumor volume, increasing survival, and maintaining the functional properties of peritumoral neurons. However, being unable to cross the blood–brain barrier (BBB), CNF1 requires injection directly into the brain, which is a very invasive administration route. Thus, to overcome this pitfall, we designed a CNF1 variant characterized by the presence of an N-terminal BBB-crossing tag. The variant was produced and we verified whether its activity was comparable to that of wild-type CNF1 in GBM cells. We investigated the signaling pathways engaged in the cell response to CNF1 variants to provide preliminary data to the subsequent studies in experimental animals. CNF1 may represent a novel avenue for GBM therapy, particularly because, besides blocking tumor growth, it also preserves the healthy surrounding tissue, maintaining its architecture and functionality. This renders CNF1 the most interesting candidate for the treatment of brain tumors, among other potentially effective bacterial toxins.

Keywords: glioblastoma; drug discovery; cytotoxic necrotizing factor type 1; protein purification; recombinant protein production

Key Contribution: We designed and produced the CNF1 variant An2-CNF1-H8 that might be endowed with the ability to cross the blood–brain barrier when intravenously injected. We characterized its cellular activity and demonstrated that the An2-CNF1-H8 variant preserves the same original activity profile of the wild-type CNF1 in different cell types, thus representing a promising therapeutic strategy for glioblastoma, also considering its beneficial effects on surrounding tissues.

---

1. Introduction

Gliomas are primary tumors occurring in the central nervous system (CNS) that arise from glial cells. Glioblastoma (GBM; World Health Organization grade IV glioma) [1] represents the most malignant form, being highly invasive and not responsive to standard treatments that include surgical resection of the tumoral mass followed by combined radio- and chemo-therapy. Therefore, the life expectancy after diagnosis is very poor [2].

We have previously shown that the *Escherichia coli* protein toxin cytotoxic necrotizing factor 1 (CNF1) is remarkably effective as an anti-neoplastic agent in a mouse model of glioma, reducing the tumoral mass, increasing the survival, and maintaining the functional properties of peritumoral neurons [3,4]. CNF1 specifically modulates Rho GTPases, thus affecting fundamental cellular processes that may induce different outcomes in different cell types. In fact, besides triggering an epithelial-mesenchymal transition in transformed epithelial cells where the movement and proliferating activity of the toxin are increased [3], CNF1 inhibits the growth of glioma cells that finally undergo death. Its detrimental activity on GBM cells is probably due to its ability to block cytodieresis in proliferating cells, leading them to senescence and death, whereas the beneficial activity on peritumoral neurons is still under investigation but seems to rely on the enhancement of synaptic plasticity and the preservation of functional attributes. These abilities point at CNF1 as a possible new strategy for the treatment of brain tumors.

However, being unable to cross the blood–brain barrier (BBB), CNF1 requires an intracerebral injection, as evidenced in all studies so far conducted using CNF1 on mouse models for CNS pathologies, including Rett syndrome (RTT) [5–7], Alzheimer's disease [8], epilepsy [9] as well as GBM [4,10,11]. Therefore, considering the invasive route of administration, the difficulty to propose the toxin as an innovative therapy prompted us to seek more tolerable delivery systems. To address this point, we have designed and produced a CNF1 variant that might be endowed with the ability to cross the BBB when intravenously (*iv*) injected (An2-CNF1-H8). Its main feature consists of the addition of an N-terminal Angiopep-2 (An2) peptide which should promote receptor-mediated transcytosis [12]. The recombinant protein was completed by the addition of a C-terminal His-tag (H8) to be used for the recombinant toxin purification. The demonstration that An2-CNF1-H8 preserves the original activity profile of the wild-type (*wt*) protein in different cell types, would also reinforce the hypothesis of CNF1 as a new candidate drug for the treatment of other disorders, such as RTT and mitochondrial encephalomyopathies. In particular, GBM cells have been the recipient for a more accurate investigation, propaedeutic to in vivo studies.

2. Results

2.1. Design and Production of a CNF1 Variant Bearing the Potentiality to Cross the BBB When Peripherally Injected

We have previously demonstrated that a carboxy terminally his-tagged version of CNF1 (CNF1-H8) can be efficiently expressed in and purified from recombinant *E. coli* BL21(DE3) cells [13]. In this study, we attempted the production of a slightly modified version of the same toxin harboring an N-terminal peptide, named Angiopep-2 (An2), and capable of interacting with low-density lipoprotein receptor-related protein-1 (LRP1). This receptor is known to be involved in the transcytosis of the brain delivery vector An2 [12]. This new version of the protein was named An2-CNF1-H8, as it still possessed a C-terminal polyhistidine tag (Figure 1A). The recombinant expressions of both CNF1-H8 and An2-CNF1-H8 were carried out at a low temperature of 15 °C, as previously reported [13]. The analysis of total cellular extracts obtained at the end of the bacterial growths indicated that an overexpression was clearly visible for both the constructs via SDS- polyacrylamide gel electrophoresis (PAGE) (Figure 1(B1)) and western blot (Figure 1(B2)). It is worth noting that the addition of the N-terminal An2 peptide led to a significantly higher production of the protein in comparison with the CNF1-H8 (Figure 1B). This outcome might be due to either a more efficient translation initiation or an increased half-life of the protein. Nevertheless, in terms of recombinant production of An2-CNF1-H8, this boost proved to be unfruitful, as the protein mainly accumulated as insoluble aggregates. This result is depicted in Figure 1C, where soluble (S) and insoluble (P) fractions, collected after cellular disruption, were analyzed via SDS-PAGE (Figure 1(C1,C3)) and western blot (Figure 1(C2,C4)). In the case of CNF1-H8 bearing bacteria, the recombinant product mainly accumulated as a soluble protein, as observable in the left part of Figure 1C. Conversely, *E. coli* synthesizing An2-CNF1-H8 suffered from the formation of inclusion bodies, distinguishable both in the Coomassie-stained gel and in western blot in the right part of Figure 1C. Despite a considerable loss of An2-CNF1-H8 as insoluble aggregates, its recovery in the cellular soluble fraction was still feasible using a previously established procedure [13].

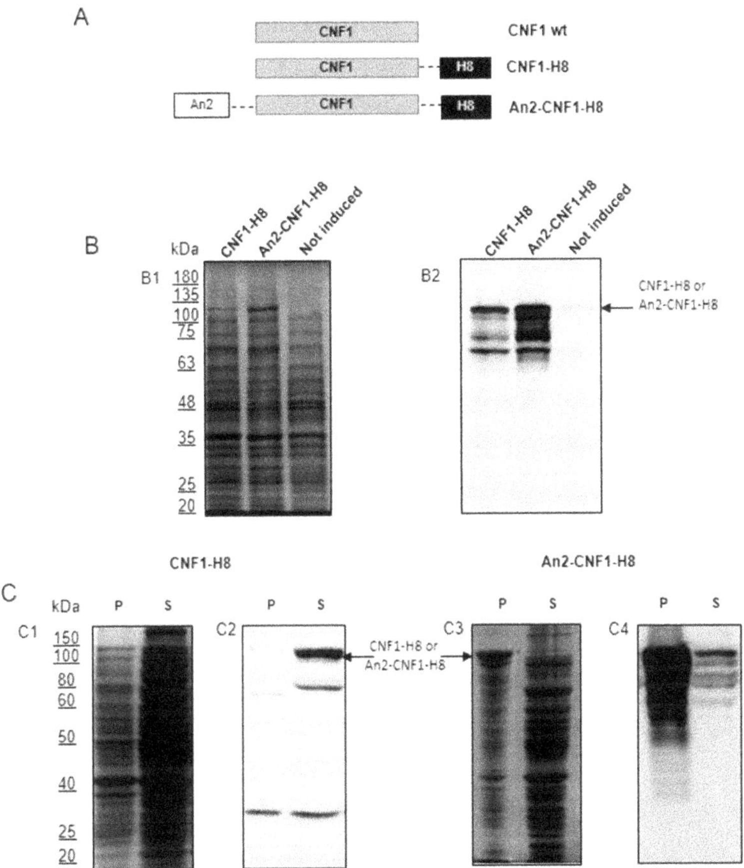

Figure 1. Design, expression, and solubility of cytotoxic necrotizing factor 1 (CNF1) variants. (**A**) Scheme of engineered CNF1 constructs presenting the following N-terminal and C-terminal tags: Angiopep-2 (An2), spaced by the GGSSRSS linker; 8xHis fused to a Tobacco Etch Virus recognition site (H8). (**B**) Recombinant protein expression levels in *E. coli* BL21 (DE3). Total cellular extracts obtained at 15 °C after 18 h induction with 0.5 mM IPTG were separated by 10% SDS-PAGE and analyzed either by Coomassie staining (**B1**) or western blot (**B2**) using a monoclonal anti-CNF1 antibody (NG8). Induced cells expressing CNF1-H8; induced cells expressing An2-CNF1-H8; not induced recombinant cells harboring pET40b-*cnf1-h8*. (**C**) Evaluation of CNF1 solubility in *E. coli* extracts. After cellular lysis, insoluble (P) and soluble (S) fractions were analyzed. The first two panels are relative to the solubility assay concerning CNF1-H8 expression carried out using SDS-PAGE (**C1** and **C3**) and western blot (**C2** and **C4**). The third and fourth images represent the SDS-PAGE and western blot relative to An2-CNF1-H8 solubility, respectively.

2.2. The An2-CNF1-H8 Variant Preserves the Original Activity Profile of wt CNF1 in HEp-2 Cells

The activity of the An2-CNF1-H8 variant was first tested on HEp-2 cells. The first set of experiments was carried out to verify whether the An2-CNF1-H8 variant could act in the same vein of CNF1. Therefore, to address this point, we used HEp-2 cells, the best characterized and responsive cell model for CNF1. CNF1-H8, and *wt* CNF1 were employed as controls in all the experiments. Once assessed that the An2-CNF1-H8 sample was more concentrated than *wt* CNF1 and CNF1-H8, as evidenced by polyacrylamide gel and western blot analysis (Figure 2A), we performed a titration experiment to find the lowest concentration inducing CNF1-like effects, i.e., multinucleation and ruffling (Figure S1). We found that the morphological effects induced by the An2-CNF1-H8 variant were comparable to that of *wt* CNF1 and obtained in the same concentration range (Figure 2B).

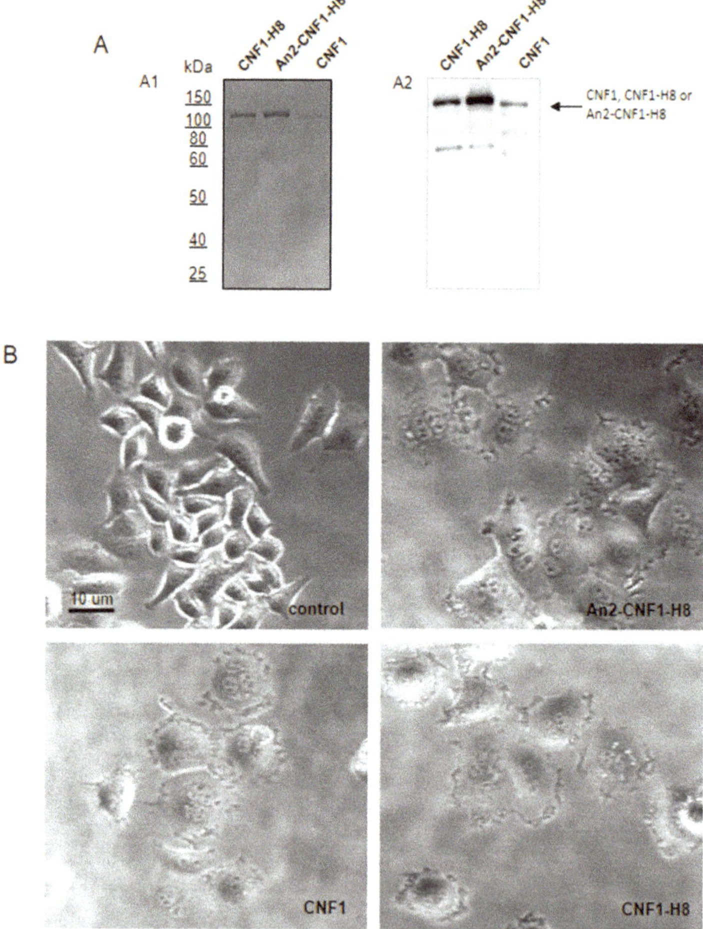

Figure 2. CNF1 variants activity. (**A**) SDS-polyacrylamide gel (**A1**) and western blot (**A2**) of CNF1 variants. (**B**) Phase-contrast micrographs of HEp-2 cells treated with CNF1 variants, showing the multinucleating/ruffling effects. Scale bar: 10 μm.

2.3. *The An2-CNF1-H8 Variant Enters the Cells in the Same Vein of wt CNF1*

To investigate whether the An2-CNF1-H8 variant binds to the same receptor engaged by the *wt* CNF1 or if it can enter via alternative pathways or receptors, we performed experiments in which we challenged the non-catalytic mutant CNF1-C866S (mCNF1, which maintains the ability to bind to the receptor but does not possess the enzymatic activity) with the CNF1 variant. As shown in Figure 3A, as the dose of mCNF1 increases, the An2-CNF1-H8 variant is inhibited, i.e., the percentage of multinucleated cells is decreased to zero, proving the engagement of the same receptor used by *wt* CNF1, at least in HEp-2 cells.

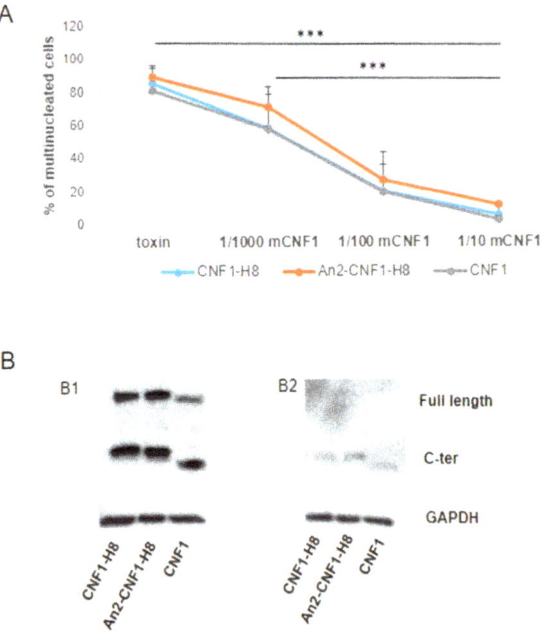

Figure 3. Entry of CNF1 variants into HEp-2 cells. (**A**) Graph showing the inhibition of variants entry by mCNF1. (**B**) Western blot analysis of HEp-2 cells treated with An2-CNF1-H8 and its controls (CNF1-H8 and *wt* CNF1) and stained with an antibody against CNF1, normalized as a function of GAPDH. Note that all the toxins are found as full length and 55 kDa C-ter portions after 24 h treatment (**B1**). Over a longer period of time, that is 48 h of treatment, washing to eliminate any remaining molecules in the culture medium, and further culture for 24 h, the full-length portions completely disappeared while the 55 kDa C-ter portions were present at a much lower concentration (**B2**). All in all, these results point at the fact that An2-CNF1-H8 follows the same route of entry of *wt* CNF1. Variants and *wt* CNF1 were used at 10^{-10} M. Data on the graph represent the mean ± SEM from at least three independent experiments. *** $p < 0.001$.

Then, to verify if the variant enters the cells via the canonical pathway of CNF1 endocytosis, being processed to release the C-terminal catalytic fragment (CNF1-C-ter) into the cytoplasm, HEp-2 cells were treated with An2-CNF1-H8 at the concentration of 10^{-9} M and lysed after 24 h. As evidenced in Figure 3(B1), the CNF1-C-ter portion of the variant was present in the cytoplasm, proving the entry through the classical CNF1 pathway. At the same time, we checked if all the administrated toxin was internalized over a longer period of time. To address this last question, we washed the cells with fresh medium after 48 h of treatment, to eliminate any remaining molecules left in the culture medium. The cells were then cultured for further 24 h before being lysed and subjected to western blot using the anti-CNF1 against the CNF1-C-ter. It should be noted that in these conditions (Figure 3(B2)) the whole CNF1 completely disappeared and the C-ter parts were present at a much lower concentration if compared to cells exposed for 24 h only. This indicates that, not only all the proteins are processed by freeing the toxin C-ter portion in the cytoplasm, but also that an amount of C-ter portion undergoes an elimination process.

2.4. Activities of the An2-CNF1-H8 Variant on the Human Brain Endothelial Cell Line HBEC-5i

The variants were built to find an inoculation system less invasive than the intracerebroventricular injection, such as the *iv* administration. Since *iv*-injected toxins need to act on endothelial cells in order to cross the BBB, we verified the activity of the CNF1 variant on the human brain endothelial cell line HBEC-5i. Interestingly, HBEC-5i cells highly express the LRP1 receptor for An2 and, therefore, we analyzed whether the variant could enter in this cellular system more easily than in the others. As reported in Figure 4, the An2-CNF1-H8 variant induces modification of the actin architecture on HBEC-5i cells comparable to that of *wt* CNF1, also in terms of fluorescence intensity (Figure S2). Moreover, competition experiments with the mutant CNF1 C866S were carried out in this endothelial cell line as in HEp-2 cells, and the results were overlapping (not shown). Therefore, our results demonstrate that An2-CNF1-H8 behaves as the *wt* CNF1 even in HBEC-5i cells. On the other hand, our preliminary in vivo experiments show that when mice were *iv* inoculated with the An2-CNF1-H8 variant, spinophilin amount increased in the hippocampus after 4 h (data not shown) as well as 7 days from inoculation (Figure S3), suggesting that the variant could pass the BBB.

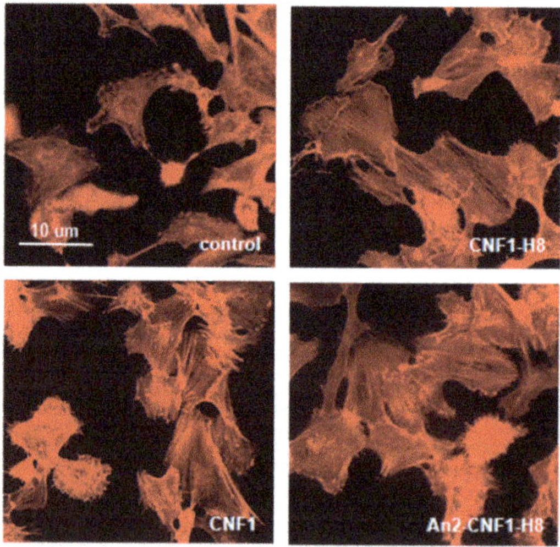

Figure 4. An2-CNF1-H8 activity on endothelial cells. Fluorescence micrographs of HBEC-5i cells treated with the An2-CNF1-H8 variant and its controls. Note the same actin architecture in cells exposed to the *wt* CNF1 and to the variants, indicating the same activity. Scale bar: 10 μm.

2.5. Activities of the An2-CNF1-H8 Variant on the Human GBM Cells

Since CNF1 has been reported to be remarkably effective as an antineoplastic agent in a mouse model of GBM, and to induce toxicity in GBM cells obtained from surgical specimens (WHO grade IV) [4,10,11], we verified whether the An2-CNF1-H8 variant could be effective on a human GBM cell line, the U87MG cells. We performed a cell growth analysis and, as shown in Figure 5A, whereas control cells underwent an increase in cell number over time, although characterized by a biphasic growth possibly due to depletion of some rate-limiting resource, cells treated with different concentrations of An2-CNF1-H8 stopped growing, starting from 24 h of exposure. The same result was obtained with the *wt* CNF1 (Figure 5B). A morphological analysis showed that the An2-CNF1-H8 variant could induce F-actin modification, in term of stress fibers and ruffles, and multinucleation. Interestingly, it caused an increase in the number of mitochondria but a reduction in their size, with respect to controls (Figure 6A). Western blot analysis showed that Tom20, a component of the translocase of the outer mitochondrial membrane (Tom), the main mitochondrial entry gate for nuclear-encoded proteins, was transiently increased following treatment with An2-CNF1-H8, with the maximum increment at 24 h. After 48 h of An2-CNF1-H8 treatment, Tom20 expression returned similar to the control level (Figure 6B). The amount of OPA-1 protein, involved in mitochondrial fusion, was not modified by the variant An2-CNF1-H8 (Figure 6B), as reported for *wt* CNF1 in other model systems [9,14]. We wondered whether the An2-CNF1-H8 variant stimulates senescence or apoptosis to cause cell number reduction. We then analyzed Bax, a pro-apoptotic protein, and found that An2-CNF1-H8 variant treatment induced an increase in Bax concentration starting from 24 h of exposure (Figure 6C), thus suggesting an involvement of apoptosis in the variant killing activity. However, we failed to observe the activation of caspase-3 (data not shown). On the other hand, western blot analysis of the senescence-associated beta-galactosidase (beta-gal) showed no significant differences (Figure 6C).

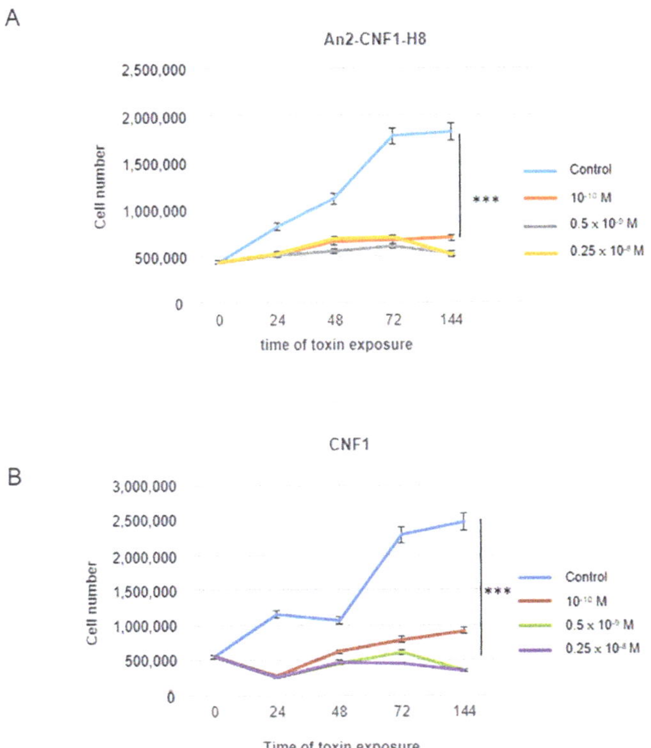

Figure 5. An2-CNF1-H8 impairs growth in U87MG GBM cells. (**A** and **B**) Graphs showing cell growth impairment by An2-CNF1-H8 (**A**) and CNF1 (**B**) on U87MG GBM cells. Note that, whereas controls cells increase in cell number in the function of time, cells treated with An2-CNF1-H8 and CNF1 stopped growing. Data on the graph represent the mean ± SEM from at least three independent experiments. *** $p < 0.001$.

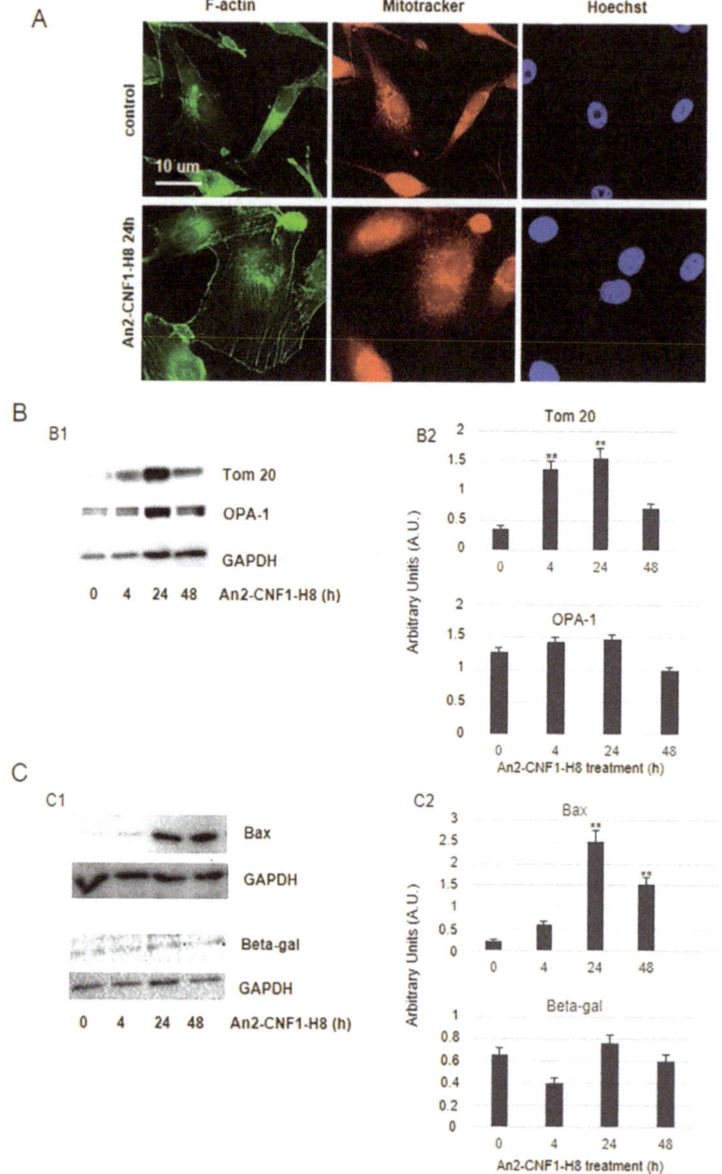

Figure 6. Mitochondrial impact and toxicity of An2-CNF1-H8 on U87MG cells. (**A**) Fluorescence micrographs of U87MG cells treated with An2-CNF1-H8 and stained with phalloidin, to detect F-actin, Mitotracker, to detect mitochondria and Hoechst 33342 to stain nuclei. Note that the variant impacts on mitochondria that appear slightly fragmented and scattered throughout the cell body with respect to controls. (**B** and **C**) Western blot of U87MG cells treated with An2-CNF1-H8 and graph analyses, showing that (**B1** and **B2**) Tom20 was transiently increased, whereas OPA-1 protein was not modified by the An2-CNF1-H8; (**C1** and **C2**) the pro-apoptotic Bax protein, involved in mitochondrial outer membrane permeabilization, was increased whereas the senescence marker beta-gal remained unvaried. Proteins were normalized as a function of GAPDH. ** $p < 0.01$. Scale bar: 10 μm.

3. Discussion

We previously reported an efficient strategy for the production and purification of the CNF1 toxin as a his-tagged recombinant protein (CNF1-H8) from *E. coli* extracts [13]. In the present manuscript, we describe the production and purification of the An2-CNF1-H8 variant, a CNF1 variant displaying an N-terminal Angiopep tag potentially endowed with the ability to allow BBB crossing.

The recombinant production of the An2-CNF1-H8 variant was approached following the protocol optimized for CNF1-H8. It turned out that the amount of recombinant An2-CNF1-H8 variant was significantly higher, but it largely accumulated as inclusion bodies, although the production was carried out at 15 °C by using only 0.5 mM inducer. Lowering of the production temperature can be a successful approach to get the soluble product when the protein has the tendency to accumulate as inclusion bodies, as in the case of CNF1-H8 [13]. However, the higher production yields achieved for An2-CNF1-H8 shifted the equilibrium towards the formation of insoluble aggregates. Nevertheless, its purification from the soluble extracts in native conditions was still feasible. Once purified, the cellular activity of the An2-CNF1-H8 variant was compared to that of *wt* CNF1 and it turned out that cellular effects completely overlapped, causing multinucleation and actin cytoskeleton changes in the same concentration ranges.

It is interesting to note that the route of entry into cells followed by the An2-CNF1-H8 variant is the same used by the *wt* CNF1. It is known that CNF1 binds to its receptor [15,16], enters endocytic vesicles by receptor-mediated endocytosis and is routed to the endosomal compartment [17] where its catalytic domain (55-kDa fragment) is transferred into the cytosol [18]. The mutant toxin CNF1 C866S, herein indicated as mCNF1, devoid of enzymatic activity [19] was able to compete for the binding to the receptor with the An2-CNF1-H8 variant inhibiting its multinucleating activity. Besides, a 55-kDa fragment was recorded into the cytosol. All these findings indicate that the variant behaves in the same way of *wt* CNF1, demonstrating that the addition of An2 and H8 at the N- and C-terminal end, respectively, did not compromise its activity on epithelial cells, although further studies on neuronal cell models should also be performed. When endothelial cells were exposed to An2-CNF1-H8, a typical CNF1-induced actin rearrangement was observed, indicating that the variant entered into this cell line and exerted its action. Since HBEC-5i cells highly express the LRP1 receptor for An2, we were expecting a better efficiency of An2-CNF1-H8 in its entry into this cell type with a consequent major effect on the targets. The observed actin rearrangement, however, was in line with what so far documented in other cell systems, underlying the necessity of further experiments to verify our hypothesis. It is interesting to note that *iv* inoculation of the An2-CNF1-H8 variant in mice can increase the amount of spinophilin in the hippocampus. This result, besides suggesting the ability of the variant to link its LRP1 receptor, represents an important step forward the demonstration of the An2-CNF1-H8 ability to cross the BBB. Further studies are needed, however, to confirm that An2-CNF1-H8 can actually pass through the barrier.

Our previous studies disclosed the potentiality of CNF1 as a novel therapeutic strategy for the treatment of CNS tumors. Vannini and coworkers reported that CNF1 possesses a cytotoxic effect on a GBM cell line as well as on primary human GBM cells, also permitting a long-term survival in a murine GBM model [10]. Besides, CNF1 reduces GBM growth and saves the function and structure of neurons [4]. Our results demonstrate that the An2-CNF1-H8 variant exerts a cell growth arrest of U87MG GBM cells in the same vein of CNF1. From a morphological point of view, it is interesting to note that An2-CNF1-H8, although inducing typical CNF1 effects consisting of F-actin polymerization and multinucleation [20], impacts on mitochondria in a different way in comparison to data reported so far. On epithelial cells and fibroblasts, CNF1 induces mitochondria elongation [21,22] due, at least in epithelial cells, to inhibition of mitochondrial fission [14]. Here, we report that, in U87MG GBM cells, both the An2-CNF1-H8 variant and CNF1 (data not shown), induced fragmentation of mitochondria and did not vary the amount of OPA1 protein, involved in mitochondria fusion. However, the number of mitochondria seems to be increased in cells challenged with the variant with respect to control cells, as also suggested by the transient increase in Tom20, a component of the outer mitochondrial membrane [23]. The dynamic of mitochondria allows the cell to respond to its ever-changing physiological conditions and in this context, fragmented mitochondria can be found in damaged or quiescent cells, associated with functional defects or early in the apoptotic pathway [24]. For this reason, we analyzed some proteins of the apoptotic as well as of the senescence pathway. Our results demonstrate that the An2-CNF1-H8 variant increased the amount of the pro-apoptotic protein Bax, although we failed to find a concomitant activation of caspase-3. On the other hand, the level of the senescence-associated marker beta-gal was almost unchanged with respect to control cells. Further studies are needed to dissect the cellular pathway engaged by the An2-CNF1-H8 variant to arrest cell growth of GBM cells. In fact, senescence may be a late phenomenon, as reported by Vannini and coworkers [10] whereas caspases may be activated later in our model. Besides, we cannot rule out the hypothesis that necroptosis can occur in U87MG GBM cells exposed to the An2-CNF1-H8 variant [24]. It is worth noting that CNF1 is being increasingly recognized as a pro-carcinogenic agent in the intestine [3,25], indicating that the toxin behaves in two different manners in the intestine and in the brain. This is not surprising since bacterial toxins, by acting as signaling modifying agents, are profoundly influenced by several factors such as, for example, different environmental and cellular/tissue conditions. In this context, it is well established that CNF1 exerts beneficial effects in mice models of Rett Syndrome and Alzheimer once injected in the brain [6,8]. On the other hand, we have very recently reported that the CNF1 ability to promote carcinogenic traits in epithelial cells is dramatically influenced by the surrounding microenvironment as well as by the cell type [3]. In all the cell types studied so far, CNF1 causes mitochondrial elongation whereas we herein report that the CNF1 variant causes mitochondria fragmentation in GBM cells, reinforcing the emerging hypothesis that the target cell type plays a pivotal role in the response to an infectious agent [26]. This last finding on mitochondria highlights a new perspective for the understanding of CNF1 activity on GBM cells.

Altogether, our results point at the An2-CNF1-H8 variant as a promising tool against GBM, preserving the same cytotoxic activities of CNF1 but being characterized by the potential ability to cross the BBB.

4. Materials and Methods

4.1. Production and Purification of CNF1 and CNF1 Variants

cnf1-h8 and *an2-cnf1-h8* genes were expressed using pET40b plasmid. The preparation of pET40b-*cnf1-h8* has been previously described [13]. For the development of pET40-*an2-cnf1-h8*, the 5′ extremity of *cnf1-h8* was replaced with a newly synthesized one using NdeI/BglII double digestion. The new DNA fragment of 188 bp was purchased from Thermo Fisher Scientific (Thermo Fisher Scientific, Waltham, MA, USA) and encodes the An2 peptide (MGTFFYGGSRGKRNNFKTEEY) coupled to a linker (GGSSRSSS) that has already been used for the production of a GFP variant fused to An2 [27].

Both An2-CNF1-H8 and CNF1-H8 were produced in *E. coli* BL21(DE3) and purified using two chromatographic steps as reported [13]. For preliminary solubility analysis, the recombinant cellular pellets were re-suspended in 5 mL/g wet cell weight of Lysis buffer (50 mM TrisHCl pH 8.0, 150 mM NaCl, 20 mM Imidazole, 15% v/v glycerol) supplemented with the complete EDTA-free protease inhibitor cocktail (Roche, Basel, Switzerland). Afterward, the suspensions were sonicated (30 s cycles, 25% amplitude with 30 sec pauses between each cycle for a total process of 20 min at 4 °C) and the soluble and insoluble fractions were separated by centrifugation (13,000 g, 4 °C, 45 min). Then, the insoluble fractions were re-suspended in an equal volume of Lysis buffer as its relative soluble fraction and the total protein profiles were studied using 10% SDS-PAGE. For western blot analysis, an anti-CNF1 antibody (NG8, Santa Cruz Biotechnology, Dallas, TX, USA, diluted 1:1000) was used following the manufacturer's instructions.

CNF1 *wt* was obtained from the *E. coli* pISS392 strain (kindly provided by V. Falbo, Istituto Superiore di Sanità, Rome, Italy), and the plasmid coding for the recombinant protein CNF1 C866S, in which the enzymatic activity on Rho GTPases is abrogated by a change of cysteine with serine at position 866 [28], was kindly provided by E. Lemichez (U627 INSERM, Nice, France). Both toxins were purified essentially as described [29], with some modifications in the procedure. All dilutions were prepared with 20 mM Tris, pH 7.4.

4.2. Cell Cultures

HEp-2 laryngeal carcinoma cells and U87MG human GBM cells were grown in Dulbecco's Modified Eagle's Medium (D-MEM, EuroClone S.p.A., Pero, Italy) enriched with 10% fetal bovine serum (FBS, EuroClone), 100 U/l mL of penicillin and 100 µg/mL of streptomycin. The cells were grown in a controlled atmosphere incubator (5% CO2, 90% humidity, temperature 37 °C). For the various steps, cells were detached with 10 mM EDTA and 0.25% trypsin in PBS, without calcium and magnesium, at pH 7.4. All treatments were performed in a microbiological safety cabinet.

For all experiments, the cells were seeded in Petri dishes or in 24-well plates, depending on the type of assay: (i) 24-well plates for all titration, inhibition, and competition experiments; (ii) 60 mm Petri dishes for western blot. Cells were seeded at a density of 2×10^4 cm^2 cells. Twenty-four h after seeding, cells were treated with the An2-CNF1-H8, CNF1-H8 and *wt* CNF1 for different times depending on the experiment considered.

HBEC-5i cerebral microvascular endothelium cells (ATCC® CRL3245™) were grown in 1% gelatin-coated culture flasks using DMEM:F12 with FBS 10% and 40 µg/mL endothelial growth supplement (Sigma–Aldrich, St. Louis, MO, USA).

4.3. CNF1 Variant Titration

Each variant was added by making successive dilutions in the culture medium, starting from a concentration of 10^{-7} M up to 10^{-14} M. After 24 h, HEp-2 cell multinucleation was estimated under a phase contrast microscope. At least 100 cells for each variant and for each dilution were counted, and the percentage of multinucleation was calculated. For HBEC-5i, since multinucleation is not evident, the cells were fixed in paraformaldehyde and labeled to highlight F-actin, which is significantly modified by CNF1, in terms of stress fibers and ruffles. All the toxins showed a similar activity ranging around 10^{-10} M.

4.4. Competition Experiments with CNF1 C866S

HEp-2 cells were put on ice for 20 min. CNF1 C866S was added to the dilutions of 1:10, 1:100, 1:1000 for 10 min on ice. An2-CNF1-H8, CNF1-H8, and *wt* CNF1 were added at 10^{-10} M. After 30 min on ice, cells were washed 3 times with fresh medium before being placed in an incubator at 37 °C for 24 h. Finally, for both HEp-2 and HBEC-5i cells, the effect of the variants was measured.

4.5. Protein Extraction of Polyacrylamide and Western Blot Gel

One or 5 µL of the different variants were run on 8% or 12% polyacrylamide gel. After the run, the proteins were stained using the brilliant blue Comassie followed by discoloration. For western blot, proteins separated by polyacrylamide gel were transferred onto Polyvinylidene Fluoride (PVDF) membrane (Bio-Rad Laboratories, Inc., Hercules, CA, USA) and an anti-CNF1 antibody recognizing the C-terminal was used. For western blot, cells were washed twice with cold PBS and then lysed with a denaturing pad composed of 50 mmol/L Tris-HCl, pH 7.5, 2 mmol/L EDTA, 100 mmol/L NaCl, 1 mmol/L Na3Vo4, 1% Nonidet P-40, 10 g/mL leupeptin, 5 g/mL aprotinin, and 10 g/mL phenyl-methyl-sulfonyl fluoride for 10 min. The lysed cells were collected with a scraper, placed in the Eppendorf, and centrifuged for 5 min at 1200 rpm. The protein concentration of the supernatants was determined with the "Bio-Rad Protein Dc" assay (Bio-Rad Laboratories). Regards in vivo experiments, tissues were lysed in the same buffer, clarified and quantified as above described. Ten µg of protein per sample were taken and subsequently solubilized by boiling in "Laemmli Sample Buffer" (Sigma–Aldrich). The proteins were separated on SDS-PAGE polyacrylamide gel [30], and subsequently electrically transferred onto a PVDF membrane (Bio-Rad Laboratories) using a Semi-dry system (Bio-Rad Laboratories). According to this procedure, the membrane in contact with the gel is placed between two absorbent paper sheets damped with the Semi-dry buffer inside the transfer apparatus. After saturating the free membrane sites with Tris-buffered saline-tween 20 (TBS-T) (20 mM Tris-HCl pH 7.5, 150 mM NaCl, 0.02% Tween 20) containing 5% skimmed milk, the same membrane was incubated overnight at 4 °C with the following antigen-specific antibodies diluted in TBS-T containing 5% skimmed milk: anti-CNF1-NG8 (Abcam, Cambridge, UK, mouse monoclonal, diluted 1:1000); anti-GAPDH (Santa Cruz, mouse monoclonal antibody, diluted 1:2000); anti-Tom20 and anti-beta-gal (Cell Signaling Technology, Inc., Danvers, MA, USA, diluted 1:1000); anti-OPA-1 (BD bioscience, diluted 1:1000); anti-Bax (Cell Signaling, diluted 1:1000); anti-spinophilin (Upstate, Lake Placid, NY, USA, diluted 1:1000).

Following washings in TBS-T, membranes were incubated with the species-specific HRP-conjugated secondary antibodies (Jackson ImmunoResearch, Ely, UK) and immune complexes were detected by chemiluminescent HRP substrates (Immobilion Western, Millipore, Burlington, MA, USA). The images were acquired through ChemiDoc (Bio-Rad Laboratories).

4.6. Cell Growth Experiments

U87MG cells were grown in 60 mm Petri dishes. Twenty-four h after seeding, cells were treated with the An2-CNF1-H8 and *wt* CNF1 at 3 different concentrations (10^{-10} M, 0.5×10^{-9}, 0.25×10^{-8}). At different time points, An2-CNF1-H8 and CNF1-treated cells were detached from the culture dish with 10 mM EDTA and 0.25% trypsin in PBS and counted by using a Neubauer chamber. Cell counts were conducted five times and at least three separate experiments were carried out.

4.7. Mouse Studies

C57BL6/J male mice (Charles River Italia), were housed under standard conditions (21/23 °C, 40/50% humidity, 12 h light-dark cycle, food and water ad libitum). All procedures were carried out in accordance with the European and National Legislation (63/2010 EU, DL 26/2014, respectively) and approved by the Italian Ministry of Health (n.469/2018-PR). Mice were treated by tail vein injection with An2-H8, CNF1-H8 (0.5×10^{-10} M in a final volume of 175 μL) or Tris. General health and behavior of each mouse were monitored for 10 days following *iv* injection. Then, mice were sacrificed, brain extracted from the skull, and brain regions rapidly dissected and frozen for further biochemical analyses.

4.8. Fluorescence Microscopy

U87MG control and treated cells were fixed and processed as previously described [21]. Fitc-phalloidin (Sigma–Aldrich) was used to stain F-actin and Mitotracker (Molecular Probes, Eugene, OR, USA) was used to stain mitochondria. Nuclei were stained with Hoechst 33258 (Sigma–Aldrich). Glass coverslips were observed with a fluorescence optical microscope Olympus BX51/BX52 and images were acquired using the program IAS 2000 (Delta System, Rome, Italy).

4.9. Statistical Analysis

Data are presented as mean ± sem. Statistical analysis was performed by one-way ANOVA. When a significant main effect was detected, Tukey's test for post-hoc comparisons was performed, with $p < 0.05$ as a threshold for significant differences.

Supplementary Materials: online at http://www.mdpi.com/2072-6651/12/5/291/s1, Figure S1: Hep-2 multinucleation, Figure S2: Fluorescence intensity, Figure S3: Spinophilin expression in hippocampal tissue.

Author Contributions: Conceptualization, A.F. (Alessia Fabbri) and C.F.; Software, Z.M.; Investigation, A.C., A.F. (Andrea Fortuna), S.L., L.R., and E.A.P.G.; Formal analysis, A.F. (Alessia Fabbri), S.L., L.R., M.L.T., and C.F.; Data Curation, A.F. (Alessia Fabbri); Writing—Original Draft Preparation, A.F. (Alessia Fabbri); Writing—Review & Editing, A.C., Z.M., E.P., E.A.P.G., M.L.T., C.F., and A.F. (Alessia Fabbri); Visualization, Z.M.; Funding Acquisition, M.L.T. and A.F. (Alessia Fabbri). All authors have read and agreed to the published version of the manuscript.

Funding: The support by the Italian parent's association "La fabbrica dei sogni 2—New developments for Rett syndrome" is kindly acknowledged. This work was also partially supported by the Association Française du Syndrome de Rett, project titled "Variante de la toxine CNF1 dotée d'une capacité à traverser la barrière ematoencephalique: une potentielle approche contre le syndrome de Rett".

Acknowledgments: The authors thank Rossella Di Nallo (ISS) for invaluable technical support.

Conflicts of Interest: The authors declare no conflict of interest.

References

1. Louis, D.N.; Perry, A.; Reifenberger, G.; von Deimling, A.; Figarella-Branger, D.; Cavenee, W.K.; Ohgaki, H.; Wiestler, O.D.; Kleihues, P.; Ellison, D.W. The 2016 world health organization classification of tumors of the central nervous system: A summary. *Acta Neuropathol.* **2016**, *131*, 803–820. [CrossRef] [PubMed]

2. Yang, K.; Niu, L.; Bai, Y.; Le, W. Glioblastoma: Targeting the autophagy in tumorigenesis. *Brain Res. Bull.* **2019**, *153*, 334–340. [CrossRef] [PubMed]

3. Fabbri, A.; Travaglione, S.; Rosadi, F.; Ballan, G.; Maroccia, Z.; Giambenedetti, M.; Guidotti, M.; Ødum, N.; Krejsgaard, T.; Fiorentini, C. The Escherichia coli protein toxin cytotoxic necrotizing factor 1 induces epithelial mesenchymal transition. *Cell. Microbiol.* **2019**, *22*, e13138. [CrossRef] [PubMed]

4. Vannini, E.; Olimpico, F.; Middei, S.; Ammassari-Teule, M.; de Graaf, E.L.; McDonnell, L.; Schmidt, G.; Fabbri, A.; Fiorentini, C.; Baroncelli, L.; et al. Electrophysiology of glioma: A Rho GTPase-activating protein reduces tumor growth and spares neuron structure and function. *Neuro. Oncol.* **2016**, *18*, 1634–1643. [CrossRef]

5. De Filippis, B.; Valenti, D.; de Bari, L.; De Rasmo, D.; Musto, M.; Fabbri, A.; Ricceri, L.; Fiorentini, C.; Laviola, G.; Vacca, R.A. Mitochondrial free radical overproduction due to respiratory chain impairment in the brain of a mouse model of Rett syndrome: Protective effect of CNF1. *Free Radic. Biol. Med.* **2015**, *83*, 167–177. [CrossRef]

6. De Filippis, B.; Fabbri, A.; Simone, D.; Canese, R.; Ricceri, L.; Malchiodi-Albedi, F.; Laviola, G.; Fiorentini, C. Modulation of RhoGTPases Improves the Behavioral Phenotype and Reverses Astrocytic Deficits in a Mouse Model of Rett Syndrome. *Neuropsychopharmacology* **2012**, *37*, 1152–1163. [CrossRef]

7. De Filippis, B.; Valenti, D.; Chiodi, V.; Ferrante, A.; de Bari, L.; Fiorentini, C.; Domenici, M.R.; Ricceri, L.; Vacca, R.A.; Fabbri, A.; et al. Modulation of Rho GTPases rescues brain mitochondrial dysfunction, cognitive deficits and aberrant synaptic plasticity in female mice modeling Rett syndrome. *Eur. Neuropsychopharmacol.* **2015**, *25*, 889–901. [CrossRef]

8. Loizzo, S.; Rimondini, R.; Travaglione, S.; Fabbri, A.; Guidotti, M.; Ferri, A.; Campana, G.; Fiorentini, C. CNF1 Increases Brain Energy Level, Counteracts Neuroinflammatory Markers and Rescues Cognitive Deficits in a Murine Model of Alzheimer's Disease. *PLoS ONE* **2013**, *8*, e65898. [CrossRef]

9. Travaglione, S.; Ballan, G.; Fortuna, A.; Ferri, A.; Guidotti, M.; Campana, G.; Fiorentini, C.; Loizzo, S. CNF1 enhances brain energy content and counteracts spontaneous epileptiform phenomena in aged DBA/2J mice. *PLoS ONE* **2015**, *10*, e0140495. [CrossRef]

10. Vannini, E.; Panighini, A.; Cerri, C.; Fabbri, A.; Lisi, S.; Pracucci, E.; Benedetto, N.; Vannozzi, R.; Fiorentini, C.; Caleo, M.; et al. The bacterial protein toxin, cytotoxic necrotizing factor 1 (CNF1) provides long-term survival in a murine glioma model. *BMC Cancer* **2014**, *14*, 449. [CrossRef]

11. Vannini, E.; Maltese, F.; Olimpico, F.; Fabbri, A.; Costa, M.; Caleo, M.; Baroncelli, L. Progression of motor deficits in glioma-bearing mice: Impact of CNF1 therapy at symptomatic stages. *Oncotarget* **2017**, *8*, 23539–23550. [CrossRef] [PubMed]

12. Demeule, M.; Currie, J.C.; Bertrand, Y.; Ché, C.; Nguyen, T.; Régina, A.; Gabathuler, R.; Castaigne, J.P.; Béliveau, R. Involvement of the low-density lipoprotein receptor-related protein in the transcytosis of the brain delivery vector Angiopep-2. *J. Neurochem.* **2008**, *106*, 1534–1544. [CrossRef] [PubMed]

13. Colarusso, A.; Caterino, M.; Fabbri, A.; Fiorentini, C.; Vergara, A.; Sica, F.; Parrilli, E.; Tutino, M.L. High yield purification and first structural characterization of the full-length bacterial toxin CNF1. *Biotechnol. Prog.* **2018**, *34*, 150–159. [CrossRef] [PubMed]

14. Travaglione, S.; Loizzo, S.; Rizza, T.; Del Brocco, A.; Ballan, G.; Guidotti, M.; Vona, R.; Di Nottia, M.; Torraco, A.; Carrozzo, R.; et al. Enhancement of mitochondrial ATP production by the *Escherichia coli* cytotoxic necrotizing factor 1. *FEBS J.* **2014**, *281*, 3473–3488. [CrossRef] [PubMed]

15. Kim, K.J.; Chung, J.W.; Kim, K.S. 67-kDa laminin receptor promotes internalization of cytotoxic necrotizing factor 1-expressing Escherichia coli K1 into human brain microvascular endothelial cells. *J. Biol. Chem.* **2005**, *280*, 1360–1368. [CrossRef] [PubMed]

16. Chung, J.W.; Hong, S.J.; Kim, K.J.; Goti, D.; Stins, M.F.; Shin, S.; Dawson, V.L.; Dawson, T.M.; Kim, K.S. 37-kDa laminin receptor precursor modulates cytotoxic necrotizing factor 1-mediated RhoA activation and bacterial uptake. *J. Biol. Chem.* **2003**, *278*, 16857–16862. [CrossRef]

17. Contamin, S.; Galmiche, A.; Doye, A.; Flatau, G.; Benmerah, A.; Boquet, P. The p21 Rho-activating toxin cytotoxic necrotizing factor 1 is endocytosed by a clathrin-independent mechanism and enters the cytosol by an acidic-dependent membrane translocation step. *Mol. Biol. Cell* **2000**, *11*, 1775–1787. [CrossRef]

18. Knust, Z.; Blumenthal, B.; Aktories, K.; Schmidt, G. Cleavage of escherichia coli cytotoxic necrotizing factor 1 is required for full biologic activity. *Infect. Immun.* **2009**, *77*, 1835–1841. [CrossRef]

19. Lerm, M.; Schmidt, G.; Goehring, U.M.; Schirmer, J.; Aktories, K. Identification of the region of Rho involved in substrate recognition by Escherichia coli cytotoxic necrotizing factor 1 (CNF1). *J. Biol. Chem.* **1999**, *274*, 28999–29004. [CrossRef]

20. Fabbri, A.; Travaglione, S.; Fiorentini, C. Escherichia coli cytotoxic necrotizing factor 1 (CNF1): Toxin biology, in vivo applications and therapeutic potential. *Toxins (Basel)* **2010**, *2*, 283–296. [CrossRef]

21. Giamboi-Miraglia, A.; Travaglione, S.; Filippini, P.; Fabbri, A.; Fiorentini, C.; Falzano, L. A multinucleating Escherichia coli cytotoxin perturbs cell cycle in cultured epithelial cells. *Toxicol. In Vitro* **2007**, *21*, 235–239. [CrossRef] [PubMed]

22. Fabbri, A.; Travaglione, S.; Maroccia, Z.; Guidotti, M.; Pierri, C.; Primiano, G.; Servidei, S.; Loizzo, S.; Fiorentini, C. The Bacterial Protein CNF1 as a Potential Therapeutic Strategy against Mitochondrial Diseases: A Pilot Study. *Int. J. Mol. Sci.* **2018**, *19*, 1825. [CrossRef] [PubMed]

23. Johnson, A.E.; Jensen, R.E. Barreling through the membrane. *Nat. Struct. Mol. Biol.* **2004**, *11*, 113–114. [CrossRef] [PubMed]

24. Westermann, B. Mitochondrial fusion and fission in cell life and death. *Nat. Rev. Mol. Cell Biol.* **2010**, *11*, 872–884. [CrossRef]

25. Buc, E.; Dubois, D.; Sauvanet, P.; Raisch, J.; Delmas, J.; Darfeuille-Michaud, A.; Pezet, D.; Bonnet, R. High Prevalence of mucosa-associated *E. coli* producing cyclomodulin and genotoxin in colon cancer. *PLoS ONE* **2013**, *8*, e56964. [CrossRef]

26. Keener, A.B. Host with the most: Targeting host cells instead of pathogens to fight infectious disease. *Nat. Med.* **2017**, *23*, 528–531. [CrossRef]

27. Serna, N.; Céspedes, M.V.; Saccardo, P.; Xu, Z.; Unzueta, U.; Álamo, P.; Pesarrodona, M.; Sánchez-Chardi, A.; Roldán, M.; Mangues, R.; et al. Rational engineering of single-chain polypeptides into protein-only, BBB-targeted nanoparticles. *Nanomedicine* **2016**, *12*, 1241–1251. [CrossRef]

28. Schmidt, G.; Selzer, J.; Lerm, M.; Aktories, K. The Rho-deamidating cytotoxic necrotizing factor 1 from Escherichia coli possesses transglutaminase activity. Cysteine 866 and histidine 881 are essential for enzyme activity. *J. Biol. Chem.* **1998**, *273*, 13669–13674. [CrossRef]

29. Falzano, L.; Fiorentini, C.; Donelli, G.; Michel, E.; Kocks, C.; Cossart, P.; Cabanié, L.; Oswald, E.; Boquet, P. Induction of phagocytic behaviour in human epithelial cells by Escherichia coli cytotoxic necrotizing factor type 1. *Mol. Microbiol.* **1993**, *9*, 1247–1254. [CrossRef]

30. Laemmli, U.K. Cleavage of structural proteins during the assembly of the head of bacteriophage T4. *Nature* **1970**, *227*, 680–685. [CrossRef]

Article

Inclusion of a Furin Cleavage Site Enhances Antitumor Efficacy against Colorectal Cancer Cells of Ribotoxin α-Sarcin- or RNase T1-Based Immunotoxins

Javier Ruiz-de-la-Herrán [1], Jaime Tomé-Amat [1,2], Rodrigo Lázaro-Gorines [1], José G. Gavilanes [1] and Javier Lacadena [1,*]

[1] Departamento de Bioquímica y Biología Molecular, Facultad de Ciencias Químicas, Universidad Complutense de Madrid, Madrid 28040, Spain; chaosrorri@hotmail.com (J.R.-d.-l.-H.); jaime.tome@upm.es (J.T.-A.); rodrigolazarogorines@ucm.es (R.L.-G.); jggavila@ucm.es (J.G.G.)

[2] Centre for Plant Biotechnology and Genomics (UPM-INIA), Universidad Politécnica de Madrid, Pozuelo de Alarcón, Madrid 28223, Spain

* Correspondence: jlacaden@ucm.es; Tel.: +34-91-394-4266

Received: 3 September 2019; Accepted: 10 October 2019; Published: 12 October 2019

Abstract: Immunotoxins are chimeric molecules that combine the specificity of an antibody to recognize and bind tumor antigens with the potency of the enzymatic activity of a toxin, thus, promoting the death of target cells. Among them, RNases-based immunotoxins have arisen as promising antitumor therapeutic agents. In this work, we describe the production and purification of two new immunoconjugates, based on RNase T1 and the fungal ribotoxin α-sarcin, with optimized properties for tumor treatment due to the inclusion of a furin cleavage site. Circular dichroism spectroscopy, ribonucleolytic activity studies, flow cytometry, fluorescence microscopy, and cell viability assays were carried out for structural and in vitro functional characterization. Our results confirm the enhanced antitumor efficiency showed by these furin-immunotoxin variants as a result of an improved release of their toxic domain to the cytosol, favoring the accessibility of both ribonucleases to their substrates. Overall, these results represent a step forward in the design of immunotoxins with optimized properties for potential therapeutic application in vivo.

Keywords: immunotoxin; ribotoxin; α-sarcin; RNase T1; furin; intracellular trafficking; colorectal cancer

Key Contribution: This work not only represents a step forward in optimizing the cytotoxic efficacy of immunotoxins based on RNase T1 and α-sarcin, but also highlights the development of an immunotoxin design platform based on these ribonucleases.

1. Introduction

Due to their high cytotoxicity, specificity, and effectivity, immunotoxins have arisen as potent and promising antitumor agents [1–3]. These chimeric proteins are composed of a target domain that specifically targets a tumor marker, fused to a toxic domain, responsible for the cytotoxicity [4,5]. Their mechanism of action involves, in a first step, high affinity binding to the tumor antigen by the targeting domain, followed by the internalization of the complex by endocytosis, and finally the release of the toxic domain promoting the death of target cells [6–8]. Over the last years, multiple evidences have been gathered demonstrating that the efficiency of immunotoxins depends on different aspects [9,10] such as the following: antibody functional affinity and specificity for the tumor antigen expressed on the cell surface, the complex internalization efficacy, the rate of the toxin release in its intracellular route, and its intrinsic specificity and potency.

Currently, most of the immunotoxins are designed as recombinant fusion proteins. Different antibody fragments and linkers have been used. Thus, immunotoxins have been engineered to achieve greater specificity and tumor labelling and enhanced antitumor efficacy by reducing size and improving tumor penetration. Regarding this field, antibody engineering has pursued the development of different new formats with improved properties that are being incorporated within immunotoxin constructs [11–14]. Furthermore, beyond antibodies, other specific molecules, such as interleukins and growth factors, have been employed as targeting domains, leading to the first FDA-approved immunotoxins indicated for the treatment of acute myeloid leukemia [15] or hairy cell leukemia [16].

Toxins from different origins, have also been employed such as *Pseudomonas* exotoxin A, Diphteria toxin, actinoporins, gelonin, and the plant toxin ricin, among others [17–26]. Interestingly, ribonucleases (RNases) have acquired a significant importance due to their ideal features for being included as immunotoxin toxic domains [27–32]. In particular, ribotoxins stand out within the family of extracellular fungal RNases, as part of the toxic domain of immunotoxins, due to their small size, high thermostability and resistance to proteases, poor immunogenicity, and especially because they are highly effective to inactivate ribosomes [33–38]. As proven by the previous results obtained within α-sarcin-based immunotoxins, α-sarcin arises as the most promising ribotoxin to be included in these antitumoral therapeutic designs [36,38–40]. Its specific ribonucleolytic activity against just one single rRNA phosphodiester bond, located at the sarcin-ricin loop (SRL) of the larger rRNA, causes protein biosynthesis inhibition and apoptosis [41–43].

Few studies, however, have been focused on improving the effectiveness of immunotoxins by modulating their intracellular pathway [44–46]. As a general mechanism, once the target domain binds to the tumor antigen and gets internalized, the antigen-immunotoxin complex is found in the early endosomes, where it can be later recycled and presented back into the cell membrane or finally degraded into lysosomes. Toxin release and endosomal escape depends then on its intrinsic features. The two main routes that are usually followed by toxins are the following: (1) the route via the Golgi apparatus or (2) direct translocation to the cytosol [1]. Therefore, intracellular toxin trafficking can be considered to be a key checkpoint for desired cytotoxic effects and regarding cytotoxic efficiency, toxin delivery to the cytosol appears as a well-stablished rate-limiting step [1,47].

In this sense, we have previously produced and characterized two immunoconjugates, IMTXA33αS and scFvA33T1, based in the ribotoxin α-sarcin or the nontoxic RNase T1, respectively, fused to the variable domains (scFv) of the monoclonal antibody A33, which recognize and bind specifically the tumor-associated antigen GPA33, overexpressed in most of colorectal cancers [30,36,48,49]. We have characterized in detail both immunoRNases, not only for their structural and functional features, but also as a model to evaluate the effect of the different toxic domains and the relationship between intracellular trafficking and immunotoxins cytotoxicity [39,40].

Briefly, the antitumoral activity differences observed between both constructs have been explained by two aspects. The exquisite specificity of the ribonucleolytic activity of α-sarcin against ribosomes [36,40] in comparison with that exhibited by RNase T1 [30,50] and the intracellular pathway followed by each toxic domain, being the latter extremely decisive [39]. On the one hand, regarding the enzymatic properties of RNase T1, it is a much less specific acid cyclizing ribonuclease, with preference for the hydrolysis of GpN bonds. Although it has the same catalytic mechanism as ribotoxins, the latter present structural differences and small modifications in their catalytic residues that make them highly specific in terms of their ribonucleolytic activity. On the other hand, α-sarcin release to the cytosol could be carried out directly from endosomes or from the retrograde pathway involving Golgi apparatus, due to its ability to interact with the acidic components of the endosomes and Golgi membranes. Conversely, RNase T1, a nontoxic RNase, with an acidic isoelectric point value (pI), is not able to interact with the acidic components of endosome or Golgi membranes. Therefore, its release into the cytosol is impaired, favoring its degradation in the lysosomes or its accumulation into the late Golgi apparatus (Figure 1) [39].

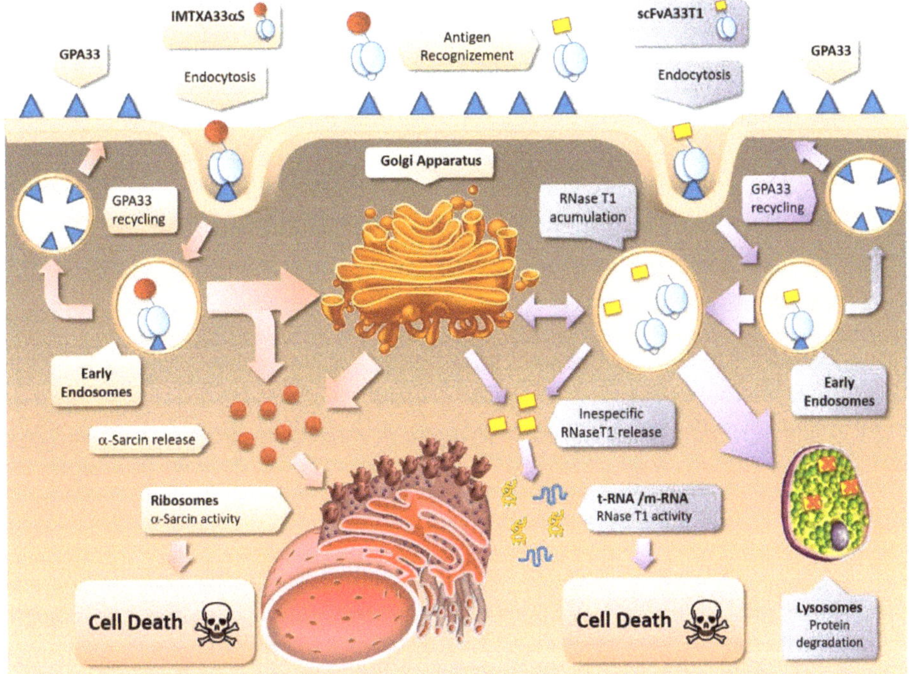

Figure 1. Scheme of the intracellular route followed by IMTXA33αS and scFvA33T1. As previously described [37], IMTXA33αS is internalized via early endosomes and follows the Golgi apparatus retrograde pathway, before α-sarcin release to the cytosol to exert its ribonucleolytic activity. Once internalized, scFvA33T1 appears also in the Golgi apparatus but mainly it is driven to lysosomes. These different pathways explain the more cytotoxic efficiency of IMTXA33αS.

In this context, some designs of immunotoxins with linkers including a specific furin cleavage site have been shown to be more cytotoxic [51–54]. This observation would confirm that intracellular processing and release of the toxic moiety is one of the key optimization spots for immunotoxin design [55–57]. Furin is a transmembrane enzyme present in the plasma membrane, the endosomes, and most notably in the trans-Golgi network [58]. It is a serine protease that belongs to the subtilisin family. Furin was known as the first mammalian proprotein-processing enzyme, exhibiting cleavage specificity for paired basic amino acid residues. Proteins were cleaved just downstream of the target sequence, canonically, Arg-X-(Arg/Lys)-Arg', with a minimal recognition cleavage site, described as Arg-X-X-Arg [59,60].

Within this idea, in this work we describe the production, purification, and in vitro functional characterization of two colorectal antitumor immunoRNases, including a furin cleavage site, based in the fungal ribotoxin α-sarcin and RNaseT1, IMTXA33furαS and scFvA33furT1, respectively (Figure 2). It is noted that our immunotoxins are usually produced in the heterologous system *P. pastoris*, a generally regarded as safe (GRAS) organism [61]. Thus, the immunotoxin secretion to the extracellular medium is driven by the α-factor signal peptide, which is finally released from the mature immunotoxin polypeptide chain by the action of Kex2 proteases. As furin belongs to this family and exhibits some of their recognition site characteristics, a minimum recognition sequence of furin has been used to avoid recognition by the Kex2 protease present in *P. pastoris* [62].

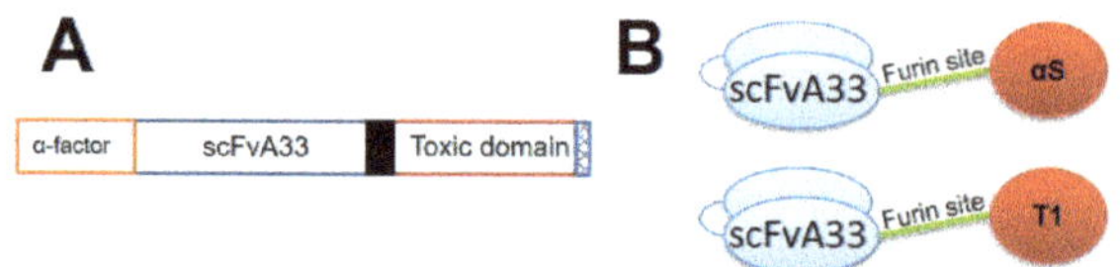

Figure 2. Schemes showing both genetic and domain structures of scFvA33-furin based immunotoxins. (**A**) Diagrammatic representation of gene construct. Both constructs include the α-factor signal peptide for secreted expression in *P. pastoris*, the anti-GPA33 scFvA33 gene (VH-linker-VL), a furin cleavage site linker (black box), the toxic domain, and the six His-tag (hatched box). The toxic domain was formed by ribotoxin α-sarcin or RNase T1. (**B**) Schematic model of the domain structure of scFvA33-furin based α-sarcin (upper) or RNase T1 (lower) immunotoxins.

2. Results

2.1. IMTXA33furαS and scFVA33furT1 Variants Were Purified as Fully Functional Immunoconjugates

Both IMTXA33furαS and scFVA33furT1 variants were successfully produced in the extracellular media of *P. pastoris* cultures after 48 h of methanol induction. Then both immunoconjugates were purified following dialysis and immobilized metal affinity chromatography (IMAC) (Figure 3A), as described in the Methods section. SDS-PAGE and western blot immunodetection, under reducing conditions, were carried out to analyze the identity and homogeneity of the purified proteins (Figure 3B). Bands of 42 and 45 kDa for scFvA33furT1 and IMTXA33furαS, respectively, corresponding to the expected theoretical molecular weight, were visualized by Coomassie blue staining and were also recognized by the anti-α-sarcin serum or anti-His-tag antibody. The final purification yield for both proteins was 1.5 and 3 mg per liter of induction medium for IMTXA33furαS and scFVA33furT1, respectively. The far-UV CD spectra for both IMTXA33furαS and scFvA33furT1 were coherent with those expected, according to the structural features of the domains that conform both immunoconjugates, and therefore compatible with water-soluble globular functional proteins. In this sense, a high content in beta sheet was observed that can be attributed to the folding described for the scFvA33, as well as to the contribution of RNase T1 and α-sarcin native conformation, which also present a high content of β-sheet and a small contribution of alpha helix [30,36] (Figure 3C).

To assess the correct functionality of these furin-variants immunoconjugates we first analyzed the ribonucleolytic activity of the two RNases included in their toxic domains. As expected, IMTXA33furαS was able to release the characteristic α-fragment, due to the specificity of its ribonucleolytic activity against the rRNA sarcin-ricin loop (SRL) present in the ribosomes (Figure 4A). Regarding scFvA33furT1, it exhibited nondistinguishable ribonucleolytic activity as compared with RNase T1 wild type or scFvA33T1 immunoconjugate when assayed in zymogram analysis against its typical substrate, poly(G), and in solution when *Torula* yeast RNA hydrolysis was measured (Figure 4B). The highly acidic feature of RNase T1 and its smaller size as compared with the immunoRNase, might explain the different intensities of the bands corresponding to scFvA33T1 and RNase T1. Thus, the incubations required to perform the assay would allow it to leak out of the gel. Both assays revealed proper functionality of the RNase moiety in the scFvA33furT1 variant. The ability of the targeting domain from both immunoconjugates to bind GPA33-positive cells SW1222 was also evaluated. As determined by flow cytometry results, both furin-variant constructs recognized GPA33-positive SW1222 cells (Figure 4C) as efficiently as IMTXA33αS and scFvA33T1, used as controls.

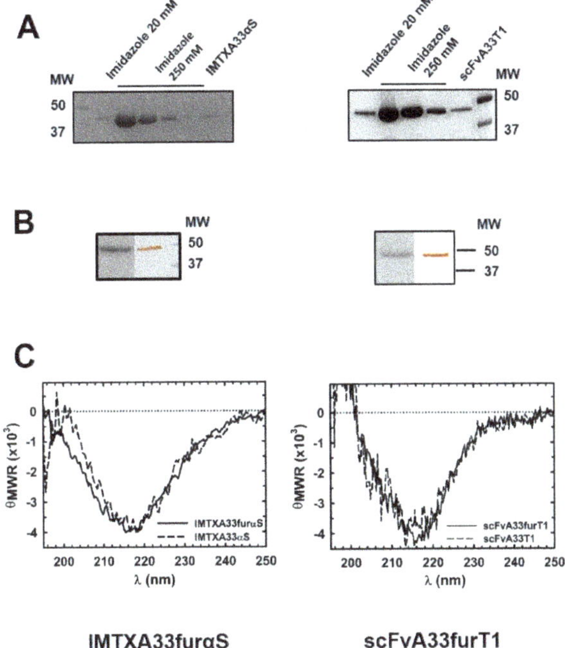

Figure 3. Production, purification, and structural characterization of furin-variant immunotoxins: IMTXA33furαS (left) or scFvA33furT1 (right). In both cases: (**A**) SDS-PAGE analysis of aliquots taken from IMAC performed to purify IMTXA33furαS or scFvA33furT1, visualized by Coomassie brilliant blue staining. (**B**) Purified IMTXA33furαS or scFvA33furT1 final fraction detection by Coomassie brilliant blue (left lane) or western blot (right lane). MW corresponds to prestained protein standards (Bio-Rad). (**C**) Far-UV circular dichroism spectra (θ_{MRW}, mean residue weight ellipticities were expressed as degree $\times$ cm^2 $\times$ dmol^{-1}) of: IMTXA33furαS (solid line) and IMTXA33αS (short dash line) (left); scFvA33furT1 (solid line) and scFvA33T1 (short dash line) (right).

Moreover, both constructs showed high structural stability and also maintained their functional integrity, when conditions mimicking a physiological context were assayed. In this sense, when purified proteins were incubated in media for at least 72 h at 37 °C, far-UV CD spectra showed that the full molecular folding of both immunotoxins were kept (Figure 5A,D). In addition, in the same conditions and for at least 72 h, the specific ribonucleolytic activity of α-sarcin (Figure 5B) and the specific antigen-binding ability of the target domain, in accordance with the flow cytometry analysis, were also preserved (Figure 5C,E).

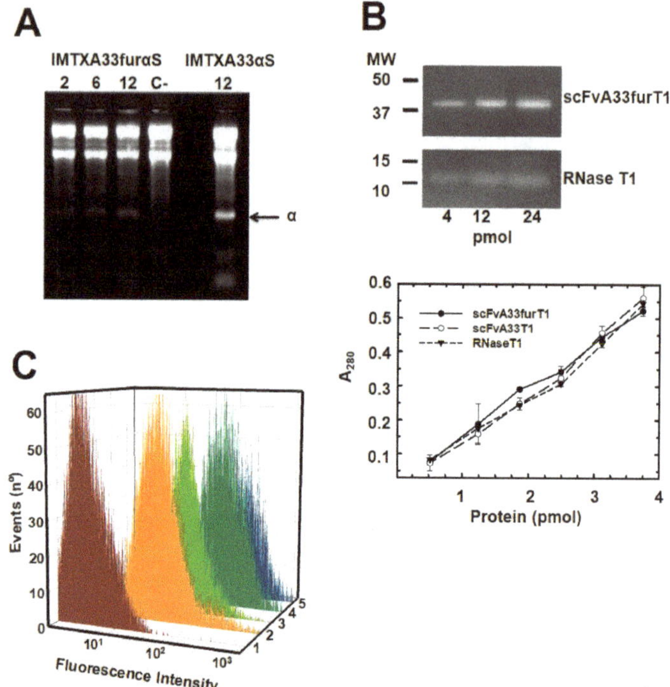

Figure 4. Functional characterization: (**A**) RNase activity assay of purified IMTXA33furαS using rabbit ribosomes as substrate. The characteristic α-fragment, as product of the specific RNase activity of α-sarcin is shown (indicated by an arrow). For IMTXA33furαS, 2, 6, and 12 pmol were assayed and 12 pmol of IMTXA33αS were used as a control. (**B**) RNase activity assays of purified scFvA33furT1. Poly(G) zymogram assay (upper panel) after SDS-PAGE of scFvA33furT1 or RNase T1. Four, 8 and 12 pmol of protein were assayed. Colorless bands were a consequence of the RNase T1 unspecific ribonucleolytic activity. MW (kDa), corresponds to electrophoretic molecular mass markers. Yeast RNA degradation assay (lower panel): The graph represents the A_{260} values versus the different amounts of scFvA33furT1 (black circles), scFvA33T1 (open circles), and RNase T1 (triangles) assayed. (**C**) Flow cytometry analysis of scFvA33furT1 (plot 2), scFvA33T1 (plot 3), IMTXA33furαS (plot 4), and (IMTXA33αS (plot 5) binding to SW1222 (GPA33-positive cells). The control curve (plot 1) corresponds to cells just treated with the anti-Histag-Alexa 488 antibody (Santa Cruz Biotechnologies, Santa Cruz, CA, USA).

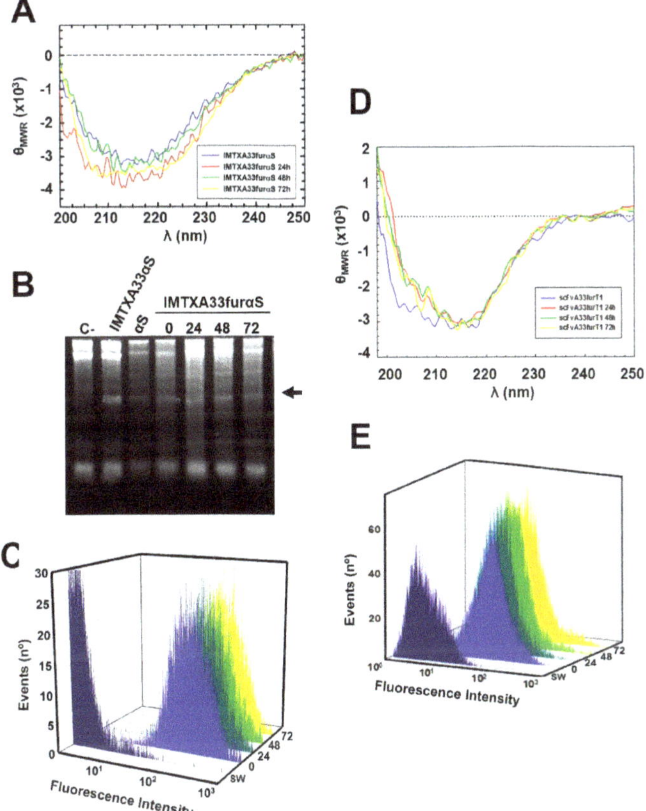

Figure 5. Stability assays of IMTXA33furαS (**A–C**) and scFvA33furT1 (**D–E**): In all cases, samples were previously incubated at 37 °C for 0, 24, 48, and 72 h. (**A,D**) Far-UV circular dichroism spectra (θ_{MRW}, mean residue weight ellipticities were expressed as degree $\times$ cm^2 $\times$ dmol^{-1}). Spectra were made with protein at 0.15 mg/mL in RPMI 1640 medium. (**B**) Specific ribonucleolytic activity against rabbit ribosomes. Samples were previously incubated at 37 °C in the absence of the substrate. The appearance of the α-fragment was indicated by an arrow. IMTXA33αS and α-sarcin wild type were used as controls. In all cases, 6 pmol of protein were assayed. (**C,E**) Binding assay o GPA33-positive SW1222 cells by flow cytometry analysis. Control curves (as in Figure 4).

2.2. IMTXA33furαS and scFVA33furT1 Follow Different Intracellular Pathway

Once it was established that the new immunoconjugate variants kept intact their antigen-specific binding capacity, as well as their ribonucleolytic activities, the next step was to analyze the effect of the furin cleavage site on the intracellular pathway followed after their internalization. In order to study this aspect, first the new constructs were labeled with Alexa 555. Both modified proteins remained fully structured, in comparison with the unlabeled versions (Figure S1), according to their far-UV CD spectra.

Internalization and intracellular localization of both labeled constructs were then studied. As shown for IMTXA33furαS (Figure 6A) and scFvA33furT1 (Figure 6B), very low colocalization was observed for both immunoconjugates with early endosomes. Although, no substantial differences were found between both constructs, interestingly, IMTXA33furαS colocalization was detected at shorter incubation times, 20 min or 2 h, than that observed for scFvA33furT1, at 5 or 24 h.

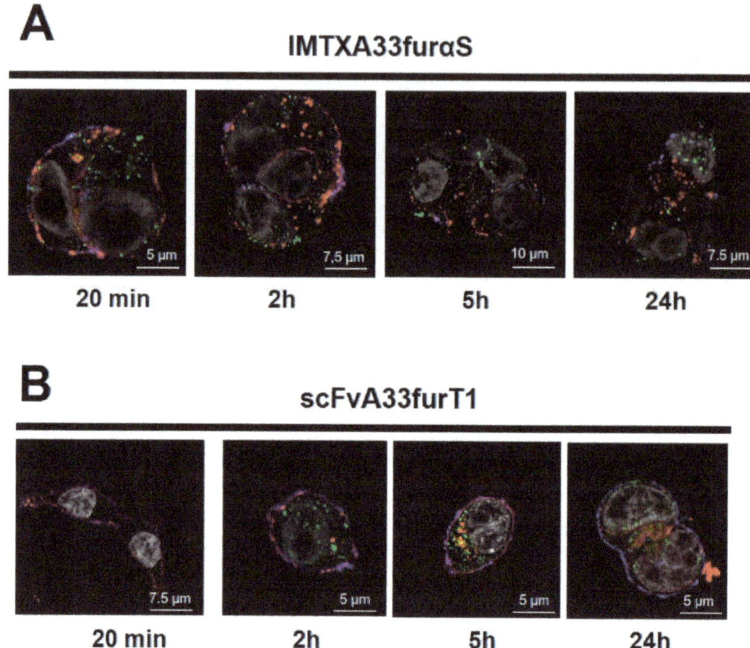

Figure 6. Endosomes colocalization analysis: Immunofluorescence confocal microscopy images obtained from cells incubated with IMTXA33furαS-555 and scFvA33furT1-555. Immunofluorescence confocal microscopy images of SW1222 cells after incubation for 20 min, 2 h, 5 h, or 24 h (from left to right) with IMTXA33furαS-555 (**A**) or scFvA33furT1-Alexa 555 (**B**). In all cases, images correspond to merging of nuclei labeled with DAPI (grey), plasmatic membrane were visualized with anti-CD44 plus GAM-Alexa 647 (blue), early endosomes were visualized with anti-EEA1 plus GAR-Alexa 488 (green), IMTXA33furαS-555 or scFvA33furT1-555 (red). The appearance of yellow dots indicates colocalization with endosomes, while violet dots correspond to colocalization with the plasmatic membrane.

As previously described for IMTXA33αS and scFvA33T1, two main intracellular pathways are followed by the toxic domain once the endosomes have been reached, via lysosomes or via Golgi-apparatus. On the one hand, for IMTXA33furαS, partial colocalization degree (22%, overlap coefficient) with Golgi apparatus at 4 h was observed, remaining at similar levels after 16 h of incubation (18%) (Figure 7A). However, colocalization with lysosomes was almost negligible. On the other hand, taken into account that for RNase T1-based immunoconjugates a significant colocalization with lysosomes was previously described [39], lysosome colocalization for scFvA33furT1 was measured. In this case, a significant colocalization was observed at 4 h, with degrees of 84%, but was dramatically decreased after 16 h (23%), and nearly disappeared when cells were treated with the antibiotic bafimolycin (<1%) (Figure 7B).

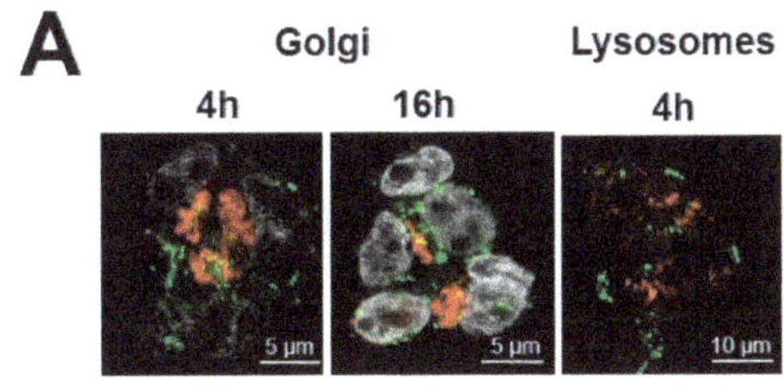

IMTXFurαS colocalization	Golgi 4h	Golgi 16h	Lysosomes 4h
Pearson's Coefficient (r)	0.192	0.18	0.03
Overlap Coefficient (%)	22.3	18.2	2.76

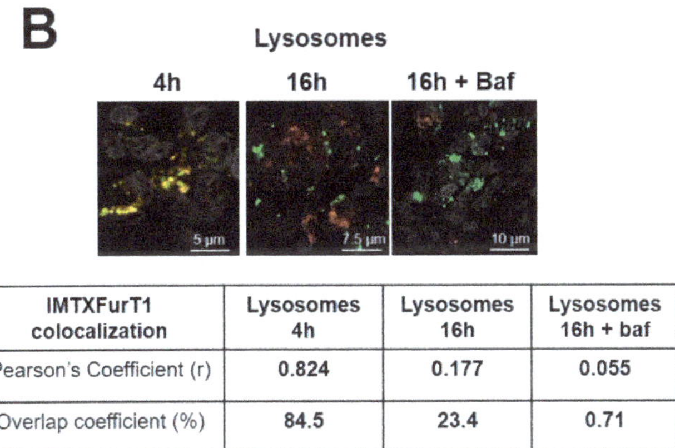

IMTXFurT1 colocalization	Lysosomes 4h	Lysosomes 16h	Lysosomes 16h + baf
Pearson's Coefficient (r)	0.824	0.177	0.055
Overlap coefficient (%)	84.5	23.4	0.71

Figure 7. Golgi apparatus and lysosomes colocalization analysis: Immunofluorescence confocal microscopy images obtained from cells incubated with IMTXA33furαS-Alexa-555 (**A**) or scFvA33furT1-Alexa-555 (**B**). (**A**) SW1222 cells were incubated for 4 or 16 h with both immunoconjugates as indicated in the figure. In all cases, images correspond to merging of nuclei labeled with DAPI (grey), Golgi with agglutinin or lysosomes with Lysotracker (red) and IMTXA33furαS-555 or scFvA33furT1-555 (green). (**B**) In the case of scFvA33T1-555, incubation was also made adding bafilomycin at 5 ng·mL^{-1}. Quantitative colocalization analysis was made as indicated in Methods. The appearance of yellow dots indicates colocalization with Golgi (**A**) or lysosomes (**B**).

The results obtained with the furin variants, compared to those described previously for the original constructions, show significant differences in the degree of colocalization with the cellular organelles involved in their intracellular pathways. On the one hand, in the case of IMTXA33furαS, although it follows the endosomas-Golgi route, just like the original construct [39], its detection in endosomes and Golgi apparatus is significantly lower, suggesting a greater or better release to the cytosol. On the other hand, for scFvA33furT1, there is a very significant increase in colocalization with the Golgi and a decrease in the case of lysosomes. As later discussed, these results would suggest that the presence of the furin cleavage site facilitates the release to the cytosol of the toxic domain of both immunoconjugates.

2.3. The Presence of the Furin Cleavage Site Significantly Increases the Cytotoxicity of Both Immunoconjugates

In vitro cytotoxicity assays were performed, to assess the effect of furin cleavage at the specific site included in these new designs. IMTXA33furαS showed an enhanced cytotoxicity with an IC_{50} two-times lower than that described for IMTXA33αS (Figure 8A). Even more interesting, cytotoxicity was increased by an additional three times when brefeldin A, that causes disassembly of Golgi-apparatus, was added (Figure 8A). However, the addition of bafilomycin that inhibits lysosomal acidification, resulted in no change in the cytotoxicity of this construction (data not shown). On the other hand, scFvA33furT1 toxicity was increased up to three times as compared to that described for scFvA33T1, being even greater when bafilomycin was added (Figure 8B). In this case no significant effect was observed with brefeldin A (data not shown).

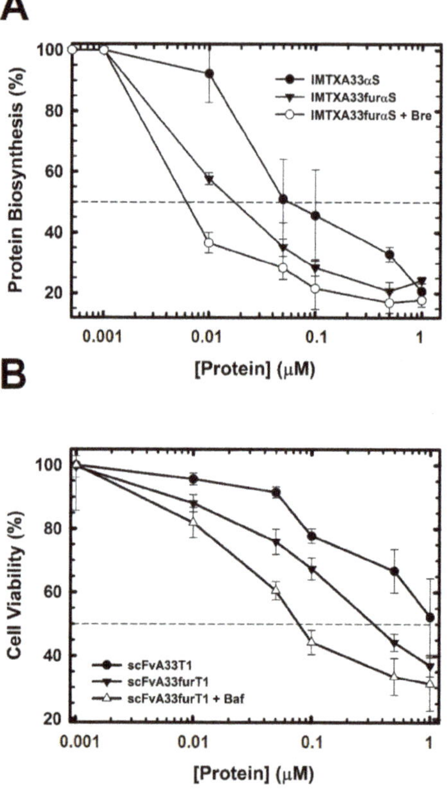

Figure 8. Cytotoxic characterization of IMTXA33furαS (**A**) and scFvA33furT1 (**B**). (**A**) Protein biosynthesis inhibition assay. SW1222 cells were incubated for 72 h with IMTXA33αS (●), IMTXA33furαS (▼), and IMTXA33aS + brefeldin A (○). (**B**) MTT viability assay of SW1222 cells incubated for 72 h with scFvA33T1 (●), scFvA33furT1 (▼), and scFvA33furT1+ bafilomycin (Δ). The 50% level of protein biosynthesis inhibition or cell viability were indicated (dashed line).

3. Discussion

We have purified and characterized two new optimized immunoconjugates, IMTXA33furαS and scFvA33furT1, which include a specific cleavage site for furin in the linker between the targeting and toxic domains of both constructs. They were purified to homogeneity from *P. pastoris* cultures, showing the expected size and folding, according to their biophysical characterization. This structural characterization yielded results which were consistent with the features previously described for their original counterparts, IMTXA33αS and scFvA33T1 [30,36], results that also were fully consistent, regarding to the nature and content of the secondary structure elements described for α-sarcin, RNase T1, and A33 scFv, their native domains [63–67]. Both constructs were also able to specifically bind to GPA33 expressed on the surface of the membrane and be internalized into SW1222 colorectal cancer cells, indicating that the targeting domain was fully functional.

Furthermore, they kept the original α-sarcin and RNase T1 ribonucleolytic activities, needed for ribosome inactivation and RNA hydrolysis, respectively. Moreover, when physiological-like conditions were analyzed, both IMTXA33αS and scFvA33T1, were stable and functional. In this sense, we could conclude that both immunoconjugates were correctly folded in terms of their structural and functional features, at least to the same extent as their original designs, IMTXA33αS and scFvA33T1.

As mentioned before, there is a close relationship between the intracellular route followed by the proteins included in the toxic domains of immunoconjugates and the antitumor efficiency of the immunotoxins [1]. Accordingly, this relationship had been previously reported for IMTXA33αS and scFvA33T1 [39]. Thus, once internalized, IMTXA33αS mainly followed the endosome-Golgi-apparatus network, whereas scFvA33T1 appeared indistinctly distributed between the lysosomes and the Golgi-apparatus. It is precisely this difference in the internalization pathway followed that led to a greater cytotoxic efficiency of IMTXA33αS [39].

Different designs of immunotoxins have been described including linkers with a furin cleavage site showing an increase in cytotoxic efficacy [51–54], confirming that intracellular processing and release of the toxic moiety is one of the key optimization spots in immunotoxin design [55–57]. Within this context, the main purpose driving this work was to evaluate the effect of including a furin linker in the previously characterized α-sarcin- and RNaseT1-based immunoconjugates [30,36,39]. The starting hypothesis was that this protease-specific site would ease the toxic moieties release and, consequently, would improve the cytotoxic efficiency of IMTXA33furαS and scFvA33furT1. Quite surprisingly, and although the results obtained for both constructs confirmed the preferred intracellular route used by each of the constructs [39], they, nevertheless, showed significant differences from what had been described for their parental designs.

The results herein presented further confirm that the cytotoxic mechanism of α-sarcin-based immunotoxins, IMTXA33furαS and IMTXA33αS, involves the endosome-Golgi intracellular pathway. The differences observed between both immunotoxins concerning the colocalization rate with the Golgi apparatus are in good agreement with their cytotoxic efficiency (Table 1). IMTXA33furαS cytotoxicity was two-fold higher than that observed for IMTXA33αS. Interestingly, the furin variant and its parent counterpart exhibited completely different behavior when traffic through the Golgi was blocked by the addition of brefeldin A. Whereas in the case of the original immunotoxin its cytotoxicity was dramatically reduced as an effect of brefeldin A, for the furin variant the cytotoxicity was significantly increased. These differences can only be explained by the presence of the furin cleavage site in the design of the immunotoxin. Furin protease activity would produce an increase in the amount of free α-sarcin that can interact with endosomal membranes and thus be released more efficiently to the cytosol, even without reaching the Golgi apparatus. It has been proven a long time ago that wild-type α-sarcin has the ability to translocate across a lipid bilayer rich in negatively charged phospholipids [68]. However, in the case of the original immunotoxin, most of the chimeric protein would reach the Golgi, and therefore the action of brefeldin A would negatively affect the release of the toxin. It must be emphasized that the inner membranes from both endosomes and Golgi apparatus have a high content of negative charges [69–72], contributing to α-sarcin translocation to the cytosol directly from endosomes or from Golgi apparatus [42,43]. Moreover, furin exert its protease activity not only in the Golgi but also in the endosomes [58]. Thus, α-sarcin release to the cytosol is facilitated by the presence of the furin cleavage site.

Table 1. Comparison of the results obtained for the furin variants vs. their originals counterparts, in relation to: level of colocalization with the Golgi apparatus and lysosomes; as well as the IC_{50} values obtained in the different cytotoxicity assays performed with just IMTXA33furαS or scFvA33furT1, or adding bafilomycin or brefeldin A, as described in Figure 8.

	IMTXA33furαS	IMTXA33αS *
Golgi colocalizationOverlap coefficient at 16 h (%)	18.2	46.3
IC_{50} (nM)	15	30–40
IC_{50} (nM) + Bafilomycin	15–20	30
IC_{50} (nM) + Brefefeldin A	5	>500
	IMTXA33furT1	IMTXA33T1 *
Lysosomes colocalizationOverlap coefficient at 16 h (%)	23.4	58.1
IC_{50} (nM)	300	>1 mM
IC_{50} (nM) + Bafilomycin	70–90	70–90
IC_{50} (nM) + Brefefeldin A	300	>1 mM

* Data from IMTXA33αS and scFVA33T1 used for comparison [39].

For scFvA33furT1, significant enhanced cytotoxicity was also observed when compared to the original design, scFvA33T1. As observed in the internalization assays, the degree of colocalization with lysosomes was significantly lower when compared to that previously described for scFVA33T1 [39] (Table 1), suggesting that most of scFvA33fur1 could follow the Golgi pathway or that the inclusion of the furin cleavage site would again favor its accumulation in endosomes. Within this, addition of bafilomycin, which impaired the recycling in the lysosomes, produced and increased in the cytotoxic effectiveness. However, the best IC_{50} value obtained for scFvA33furT1 (70–90 nM) after incubation with bafilomycin did not improve the cytotoxic effectiveness obtained by its original equivalent in the presence of the same antibiotic. Regarding this, it must be noted that the acidic pI value of RNase T1 [50,73] hamper its interaction with the endosome or Golgi membranes. Even if the lysosome pathway was partially blocked, the release into the cytosol was impaired. In fact, there is not any published account of RNase T1 being able to translocate a biological membrane on its own.

In summary, our results confirm the different routes followed by α-sarcin and RNaseT1 when included as part of the toxic domain of an immunotoxin and how we can take advantage of their features including a furin cleavage site in its design. Both furin-variant immunoconjugates exhibit enhanced antitumor effectiveness than that described for their original parents. Therefore, in vivo assays with IMTXA33furαS, exhibiting increased antitumor activity than that based in RNase T1, must be addressed.

4. Conclusions

This work not only represents a step forward in optimizing the cytotoxic efficacy of immunotoxins based on α-sarcin and RNase T1, but also highlights the development of an immunotoxin design platform based on these ribonucleases, including new designs with different specificities, with monomeric or trimeric formats [74,75], or including the use of non-immunogenic variants of α-sarcin [38]. The combination of all these optimization approaches will represent a very important boost for its application in the clinic.

5. Materials and Methods

5.1. Plasmid Design

Plasmids encoding IMTXA33αS and scFvA33T1 were previously obtained [30,36]. PCR was used to amplify the cDNA sequences of interest, using the necessary oligos to incorporate the sequence coding for the furin cleavage site between the target and the toxic domains, and also the restriction sites needed for cloning. Once purified, the resultant IMTXA33furαS and scFvA33furT1 cDNAs were cloned in pPICZαA (Invitrogen), for their expression in the methylotrophic yeast *P. pastoris KM71*. To facilitate protein detection and purification, as in their original counterparts, a 6 His-tag was included at the C-terminus of both immunotoxins (Figure 2). The expression vectors were amplified in *E. coli DH5αF′* and subsequently sequenced, using the DNA sequencing service of Universidad Complutense's Genomics Unit.

5.2. Protein Production and Purification

Electrocompetent *P. pastoris KM71* cells were used to electroporate 5–10 µg of the corresponding linearized plasmid previously cleaved with *Pme* I. A Bio-Rad Gene pulser apparatus (Bio-Rad, Berkeley, CA, USA) was used for this purpose. Multiple independent clones were selected with different amounts of zeocin (100, 400, or 750 µg/mL) and assayed to find the most productive colonies. To carry out these screenings, cells were grown in BMGY in 24-well plates at 30 °C for 24 h, were harvested afterwards, and finally suspended in BMMY. For induction of protein production, the cuture was shaked at 25 °C and 200 rpm, as previously described [30,36,39].

Once the induction was completed, the secretion to the extracellular media of the proteins of interest was analyzed by 0.1% (*w/v*) sodium dodecyl sulfate (SDS)-15% (*w/v*) polyacrylamide gel electrophoresis (PAGE) and western blot. In this sense, for the specific detection of the toxic domain by western blot, a rabbit anti-α-sarcin serum was used. In addition, an anti-histidine tag antibody was used to check integrity of the purified proteins. Large-scale production of both immunotoxins was carried out by addition of 25 mL of preinoculum to a 2 L baffled flasks containing 350 mL of BMGY. Then the culture was incubated with shaking for 16 h at 30 °C and 220 rpm. Cells were recollected by soft centrifugation at room temperature and resuspended again in 200 mL of BMMY. For induction of protein production, cells were incubated at 25 °C, 250 rpm shaking, for 48 h, supplementing the culture with methanol every 24 h. Once finalized, dialysis of the extracellular medium containing the proteins of interest was carried out, using a 50 mM sodium phosphate buffer with 0.1 M NaCl, pH 7.5.

IMTXA33furαS and scFvA33furT1 were purified from the dialyzed extracellular medium by IMAC. A Ni^{2+}-NTA agarose column was used for this purpose (HisTrap™ FF Columns, GE Healthcare, Fairfield, CT, USA). The extracellular media was applied to the column at 1 mL/min. The flow rate was controlled by a peristaltic pump. Two consecutive washes were done with the dialysis buffer and with the same but adding 20 mM imidazole, before elution of the proteins of interest, by using the same buffer but containing 250 mM imidazole. Finally, the different aliquots containing the purified proteins were collected and dialyzed again against the dialysis buffer.

5.3. Biophysical Characterization

Absorbance measurements were performed on an Uvikon 930 spectrophotometer (Kontron). As described before [76], far-UV circular dichroism (CD) spectra were carried out using a Jasco 715 spectropolarimeter. Proteins dissolved in PBS were prepared at 0.15 mg/mL. Cells of 0.1 cm optical path were used. Four spectra were averaged to obtain the final data. For stability assays, spectra were obtained after previous incubation of both constructs at 37 °C for 0, 24, 36, 48, or 72h in RPMI 1640.

5.4. Ribonucleolytic Activity Assays

The specific ribonucleolytic activity of α-sarcin, included in the toxic domain of IMTXA33furαS, was followed as described before [73,77]. For detection of the release of the characteristic 400 nt rRNA, namely α-fragment, we used as substrate ribosomes from a rabbit cell-free reticulocyte lysate. Briefly, the lysate was diluted three-fold in 40 mM Tris-HCl buffer, pH 7.5, containing 40 mM KCl and 10 mM EDTA. Then, 50 µL aliquots were taken of this dilution (5–6 pmol of ribosomes approximately) and incubated for 15 min at room temperature with different amounts of the proteins to be assayed. Then, the reaction was finished by adding 250 µL of 50 mM Tris-HCl, pH 7.4, 0.5% (*w/v*) SDS, followed by briefly vortex. Subsequently, RNA phenol/chloroform extraction was carried out. The RNA pellet obtained by the addition of isopropanol to the aqueous phase, was then washed with 70% (*v/v*) ethanol, dried exhaustively, and finally resuspended in 10 µL of DEPC H_2O. The presence of α-fragment in the samples was detected by ethidium bromide staining after electrophoresis, on denaturing 2% agarose gels. For quantification of the bands we used the Gel Doc XR Imaging System and Quantity One 1-D analysis software (Bio-Rad, Berkeley, CA, USA). Original IMTXA33αS [36] was used as the control.

The ribonucleolytic activity of RNase T1, included in scFvA33furT1, was analyzed by two parallel assays. First, RNA degradation in solution was tested by measurement of *Torula* yeast RNA (Sigma, type VI; Sigma-aldrich, St. Louis, MI, USA) hydrolysis [73]. In this assay the absorbance values obtained are proportional to the amount of soluble small oligonucleotides resulting from the RNasa activity on RNA larger size fragments contained in the substrate sample. The results obtained for the RNA samples analyzed (2 mg/mL) without protein were used to consider nonenzymatic RNA degradation, while original scFvA33T1 [30] and free RNase T1 were used as controls. In the second assay, the activity of purified scFvA33furT1 and RNase T1 was evaluated following the zymogram method [78]. Briefly, samples were applied into SDS-PAGE in 15% polyacrylamide gels, containing 0.1% (*w/v*) SDS and embedded with 0.3 mg/mL poly(G) homopolyribonucleotide [79]. Proteins exhibiting RNase activity were visualized as colorless bands, whereas the rest of the gel appeared colored.

5.5. Cell Lines Culture

Colon carcinoma SW1222 cells were used as model for the GPA33-positive cellular line [30,36,39]. The cultures were grown as described [30], in RPMI 1640 medium (Sigma-aldrich, St. Louis, MI, USA), supplemented with glutamine (300 mg/mL), containing 50 U/mL of penicillin and 50 mg/mL of streptomycin. Finally, the media was supplemented with 10% fetal bovine serum (FBS). Incubation of the cells was carried out at 37 °C in a humidified atmosphere (CO_2:air, 1:19, *v:v*). Trypsinization was routinely done for harvesting and propagation of the cultures. A hemocytometer was used to count the cells used in all assays described.

5.6. Flow Cytometry Studies

Trypsinized cells were distributed into different aliquots, containing 3×10^5 cells/mL, and washed several times with PBS 0.1% (*w/v*) BSA containing 0.02% (*w/v*) sodium azide. The samples were incubated with gentle shaking for 1 h at room temperature, with 1 µM of IMTXA33furαS or scFvA33furT1, using their original counterparts as positive controls. A second incubation, adding anti-His-Alexa488 (Sigma-aldrich, St. Louis, MI, USA) diluted 1/100, was carried out in the dark. When necessary, between the different steps, cells were collected by centrifugation (1200× *g*, 4 °C, 10 min) and then washed with PBS for several times. Flow cytometry acquisition was done on a FACScan (Becton Dickinson, Becton Dickinson, NJ, USA) and data were obtained using the WinMDI software. For stability assays, IMTXA33furαS and scFvA33furT1 were previously incubated at 37 °C for different times (0, 24 or 72 h).

5.7. Fluorescence Microscopy

To carry out these assays, both immunotoxins were first labeled. As previously described [37], using the Alexa Fluor 555 Protein Labeling Kit (Invitrogen, Carlsbad, CA, USA). Both labeled conjugates (IMTXA33furαS-555 and scFvA33furT1-555) were purified and characterized to ensure the preservation of their structural and functional features.

SW1222 cells were first trypsinized, seeded at 8×10^5 cells/well using cover-glasses, and incubated overnight at 37 °C. For treatment, IMTXA33furαS-555 or scFvA33furT1-555, at 25 µg/mL, were added to cells for the following different periods of time: from 20 min to 24 h for endosomes colocalization assays, and 4 or 16 h when lysosomes/Golgi colocalization assays were performed. Confirmation of colocalization with lysosomes was also studied by the addition of bafilomycin at 5 ng/mL, an antibiotic that inhibits lysosomal acidification [80], in combination with the immunoRNase. To observe the plasmatic membrane, incubation with anti-CD44 mAb [81] were performed. After removing the medium, the cells were fixed for 15 min with PBS containing 3% (*v/v*) p-formaldehyde and incubated for 15 min in PBS containing 50 mM ammonium chloride. Cells were permeabilized by the addition of digitonine at 0.01% (*w/v*) in PBS and incubation for 30 min, followed by another incubation in PBS containing 1.0% (*w/v*) BSA for 1 h. Different probes were used for organelle labeling as follows: an Ab targeting protein EAA1, present in the early endosomes [82]; lysosomes were labeled with Lysotracker (Life Technologies, Carlsbad, CA, USA), a fluorescence acidotropic specific probe [83]; and Golgi apparatus with Wheat Germ Agglutinin (Life Technologies) that binds to the highly abundant sialic acid present in the Golgi membranes [84]. Donkey anti-mouse Alexa 647 (DAM-Alexa 647) or goat anti-rabbit Alexa 488 (GAR-Alexa 488) were also added as secondary antibodies. For nuclei labeling, 10 µL of Prolong Gold-DAPI (Life Technologies) were added. Incubations were carried out at room temperature and the final samples were kept at 4 °C. To obtain the corresponding images, a Leica TCS SP2 confocal microscope was used, followed by analysis with the LCS lite software. Images presented correspond to the optical planes concerning the internal content of the cells selected for analysis. For each sample, ten different images were taken along the Z-axis, covering all the cells, from the basal zone to the apical one, Images shown correspond to the internal content of the cells, in particular to slices 4 to 6. ImageJ software was used to performed colocalization quantification, using two coefficients, the Pearson's correlation coefficient and, as described by Mander, the overlap coefficient. Both coefficients referred to the correlation between the intensity distribution of the channels and the true degree of colocalization, respectively [85].

5.8. MTT Viability Assay

Cell viability was evaluated by using the MTT-Cell Proliferation Kit I (Roche, Basel, Switzerland) as previously described [30,39]. Briefly, 5×10^3 trypsinized cells/well were seeded and then incubated for 24 h at 37 °C. The medium was then removed and scFvA33furT1 or scFvA33T1 were added at different concentrations in 200 µL final volume. Samples where incubated for 96 h, followed by another incubation with MTT at 0.5 mg/mL during 4 h at 37 °C. Finally, the solubilization buffer was added and the viability was determined in terms of A_{595} nm, whereas the higher A_{595} values were in correspondence with increasing amounts of viable cells. Cells incubated only with medium, in the absence of the protein, were taken as 100% viability. If necessary, samples including bafilomycin at 5 ng/mL, together with the protein were also done. Control only with bafilomycin, was included to evaluate potential drug related toxicity. The results shown correspond to the average of four independent assays.

5.9. Protein Biosynthesis Inhibition

To evaluate the cytotoxicity of ribotoxins the protein biosynthesis inhibition assay is routinely used [41–43]. Briefly, cells were seeded into 96-well plates (1×10^4 cells/well) in culture medium and kept under standard culture conditions for 24 h. Different concentrations of the different immunoconjugates, in 200 µL of free-FBS fresh medium were added to the cells. After 72 h of incubation at 37 °C, the medium was removed, followed by replacement with a fresh one containing 1 mCi per well of L-[4-C-3H]-Leucine (166 Ci/mmol; GE Healthcare, Fairfield, CT, USA). The samples were incubated again for 6 h, the medium was eliminated, and cells were then fixed with 5.0% (*w/v*) trichloroacetic acid. Several washed with cold ethanol were done, before the pellet was finally dissolved in 200 µL of 0.1 M NaOH containing 0.1% SDS. A Beckman LS3801 liquid scintillation counter was used to measure its radioactivity. Cytotoxicity was calculated in terms of IC_{50} values (namely, protein concentration needed to produce 50% protein synthesis inhibition), expressed as the percentage of the radioactivity incorporated in the assay. Three independent replicates per two assays were performed to average the IC_{50} values. When required, brefeldin A, that causes disassembly of Golgi-apparatus by altering vesicular trafficking [86], was added at 1 ng/mL together with the immunotoxins. A control, just with brefeldin A without the proteins, was performed to evaluate drug related toxicity.

Supplementary Materials: The following are available online at http://www.mdpi.com/2072-6651/11/10/593/s1, Figure S1: Structural characterization of Alexa-555 labelled furin-variant immunotoxins.

Author Contributions: Conceptualization, J.L., J.R.-d.-l.-H., and J.T.-A.; methodology, J.R.-d.-l.-H., J.T.-A., and R.L.-G.; formal analysis and investigation, J.R.-d.-l.-H., J.T.-A., R.L.-G., J.G.G., and J.L.; writing—original draft preparation, J.L.; writing—review and editing, J.L., J.R.-d.-l.-H., J.T.-A., R.L.-G., J.G.G.

Funding: This research was funded by the Universidad Complutense de Madrid, grant numbers PR41/17-21004 and PR75/18-21563.

Acknowledgments: We very much appreciate A. Martínez del Pozo for critically reading the manuscript. JTA was granted from Instituto de Salud Carlos III (ISCIII) co-founded by FEDER Thematic Networks and Cooperative Research Centers: ARADYAL (RD16/0006/0003, RD16/0006/0013, RD16/0006/0015).

Conflicts of Interest: The authors declare no conflict of interest. The funders had no role in the design of the study; in the collection, analyses, or interpretation of data; in the writing of the manuscript, or in the decision to publish the results".

References

1. Alewine, C.; Hassan, R.; Pastan, I. Advances in anticancer immunotoxin therapy. *Oncologist* **2015**, *20*, 176–185. [CrossRef] [PubMed]
2. Kavousipour, S.; Khademi, F.; Zamani, M.; Vakili, B.; Mokarram, P. Novel biotechnology approaches in colorectal cancer diagnosis and therapy. *Biotechnol. Lett.* **2017**, *39*, 785–803. [CrossRef] [PubMed]
3. Nasiri, H.; Valedkarimi, Z.; Aghebati-Maleki, L.; Majidi, J. Antibody-drug conjugates: Promising and efficient tools for targeted cancer therapy. *J. Cell. Physiol.* **2018**, *233*, 6441–6457. [CrossRef]
4. Frankel, A.E.; Woo, J.H.; Neville, D.M. Immunotoxins. In *Principles of Cancer Biotherapy*; Oldham, R.K., Dillman, R.O., Eds.; Springer: Dordrecht, The Netherlands, 2009; pp. 407–449.
5. Madhumathi, J.; Verma, R.S. Therapeutic targets and recent advances in protein immunotoxins. *Curr. Opin. Microbiol.* **2012**, *15*, 300–309. [CrossRef]
6. Pastan, I.; Hassan, R.; FitzGerald, D.J.; Kreitman, R.J. Immunotoxin treatment of cancer. *Annu. Rev. Med.* **2007**, *58*, 221–237. [CrossRef] [PubMed]
7. Kreitman, R.J. Recombinant immunotoxins containing truncated bacterial toxins for the treatment of hematologic malignancies. *BioDrugs* **2009**, *23*, 1–13. [CrossRef] [PubMed]
8. King, E.M.; Mazor, R.; Cuburu, N.; Pastan, I. Low-dose methotrexate prevents primary and secondary humoral immune responses and induces immune tolerance to a recombinant immunotoxin. *J. Immunol.* **2018**, *200*, 2038–2045. [CrossRef] [PubMed]
9. Becker, N.; Benhar, I. Antibody-Based Immunotoxins for the Treatment of Cancer. *Antibodies* **2012**, *1*, 39–69. [CrossRef]
10. Weldon, J.E.; Xiang, L.; Zhang, J.; Beers, R.; Walker, D.A.; Onda, M.; Hassan, R.; Pastan, I. A recombinant immunotoxin against the tumor-associated antigen mesothelin reengineered for high activity, low off-target toxicity, and reduced antigenicity. *Mol. Cancer Ther.* **2013**, *12*, 48–57. [CrossRef] [PubMed]
11. Muyldermans, S. Nanobodies: Natural Single-Domain Antibodies. *Annu. Rev. Biochem.* **2013**, *82*, 775–797. [CrossRef] [PubMed]
12. Ecker, D.M.; Jones, S.D.; Levine, H.L. The therapeutic monoclonal antibody market. *mAbs* **2015**, *1*, 9–14. [CrossRef] [PubMed]
13. Álvarez-Cienfuegos, A.; Nuñez-Prado, N.; Compte, M.; Cuesta, A.M.; Blanco-Toribio, A.; Harwood, S.L.; Villate, M.; Merino, N.; Bonet, J.; Navarro, R.; et al. Intramolecular trimerization, a novel strategy for making multispecific antibodies with controlled orientation of the antigen binding domains. *Sci. Rep.* **2016**, *6*, 28643. [CrossRef] [PubMed]
14. Kimiz-Gebologlu, I.; Gulce-Iz, S.; Biray-Avci, C. Monoclonal antibodies in cancer immunotherapy. *Mol. Biol. Rep.* **2018**, *45*, 2935–2940. [CrossRef] [PubMed]

15. Appelbaum, F.R.; Bernstein, I.D. Gemtuzumab ozogamicin for acute myeloid leukemia. *Blood* **2017**, *130*, 2373–2376. [CrossRef] [PubMed]

16. Kreitman, R.J.; Dearden, C.; Zinzani, P.L.; Delgado, J.; Karlin, L.; Robak, T.; Gladstone, D.E.; le Coutre, P.; Dietrich, S.; Gotic, M.; et al. Moxetumomab pasudotox in relapsed/refractory hairy cell leukemia. *Leukemia* **2018**, *32*, 1768–1777. [CrossRef]

17. Avila, A.D.; de Acosta, C.M.; Lage, A. A new immunotoxin built by linking a hemolytic toxin to a monoclonal antibody specific for immature T lymphocytes. *Int. J. Cancer* **1988**, *42*, 568–571. [CrossRef] [PubMed]

18. Foss, F.M.; Saleh, M.N.; Krueger, J.G.; Nichols, J.C.; Murphy, J.R. Diphtheria toxin fusion proteins. *Curr. Top. Microbiol. Immunol.* **1988**, *234*, 663–681.

19. LeMaistre, C.F.; Saleh, M.N.; Kuzel, T.M.; Foss, F.; Platanias, L.C.; Schwartz, G.; Ratain, M.; Rook, A.; Freytes, C.O.; Craig, F.; et al. Phase I Trial of a Ligand Fusion-Protein (DAB389IL-2) in Lymphomas Expressing the Receptor for Interkeukin-2. *Blood* **1998**, *91*, 399–405.

20. O'Toole, J.E.; Esseltine, D.; Lynch, T.J.; Lambert, J.M.; Grossbard, M.L. Clinical trials with blocked ricin immunotoxins. *Curr. Top. Microbiol. Immunol.* **1998**, *234*, 35–56.

21. Schnell, R.; Vitetta, E.; Schindler, J.; Barth, S.; Winkler, U.; Borchmann, P.; Hansmann, M.L.; Diehl, V.; Ghetie, V.; Engert, A. Clinical trials with an anti-CD25 ricin A-chain experimental and immunotoxin (RFT5-SMPT-dgA) in Hodgkin's lymphomaLeuk. *Lymphoma* **1998**, *30*, 525–537. [CrossRef]

22. García-Ortega, L.; Alegre-Cebollada, J.; García-Linares, S.; Bruix, M.; Martínez-Del-Pozo, A.; Gavilanes, J.G. The behavior of sea anemone actinoporins at the water-membrane interface. *Biochim. Biophys. Acta* **2011**, *1808*, 2275–2288. [CrossRef] [PubMed]

23. Pirie, C.M.; Hackel, B.J.; Rosenblum, M.G.; Wittrup, K.D. Convergent potency of internalized gelonin immunotoxins across varied cell lines, antigens, and targeting moieties. *J. Biol. Chem.* **2011**, *286*, 4165–4172. [CrossRef] [PubMed]

24. Wang, Z.; Duran-Struuck, R.; Crepeau, R.; Matar, A.; Hanekamp, I.; Srinivasan, S.; Neville, D.M., Jr.; Sachs, D.H.; Huang, C.A. Development of a diphtheria toxin based antiporcine CD3 recombinant immunotoxin. *Bioconjug. Chem.* **2011**, *22*, 2014–2020. [CrossRef] [PubMed]

25. Liu, X.F.; FitzGerald, D.J.; Pastan, I. The insulin receptor negatively regulates the action of *Pseudomonas* toxin-based immunotoxins and native *Pseudomonas* toxin. *Cancer Res.* **2013**, *73*, 2281–2288. [CrossRef] [PubMed]

26. Rivera-de-Torre, E.; Palacios-Ortega, J.; Gavilanes, J.G.; Martínez-del-Pozo, Á.; García-Linares, S. Pore-Forming Proteins from Cnidarians and Arachnids as Potential Biotechnological Tools. *Toxins (Basel)* **2019**, *11*, 370. [CrossRef] [PubMed]

27. De Lorenzo, C.; D'Alessio, G. From ImmunoToxins to ImmunoRNases. *Curr. Pharm. Biotechnol.* **2008**, *9*, 210–214. [CrossRef] [PubMed]

28. Balandin, T.G.; Edelweiss, E.; Andronova, N.V.; Treshalina, E.M.; Sapozhnikov, A.M.; Deyev, S.M. Antitumor activity and toxicity of anti-HER2 immunoRNase scFv 4D5-dibarnase in mice bearing human breast cancer xenografts. *Investig. New Drugs* **2011**, *29*, 22–32. [CrossRef] [PubMed]

29. Borriello, M.; Laccetti, P.; Terrazzano, G.; D'Alessio, G.; De Lorenzo, C. A novel fully human antitumor immunoRNase targeting ErbB2-positive tumors. *Br. J. Cancer* **2011**, *104*, 1716–1723. [CrossRef]

30. Tomé-Amat, J.; Menéndez-Méndez, A.; García-Ortega, L.; Batt, C.A.; Oñaderra, M.; Martínez-del-Pozo, A.; Gavilanes, J.G.; Lacadena, J. Production and characterization of scFvA33T1, an immunoRNase targeting colon cancer cells. *FEBS J.* **2012**, *279*, 3022–3032. [CrossRef]

31. Shirmann, T.; Frenzel, A.; Linden, L.; Stelte-Ludwig, B.; Willuda, J.; Harrenga, A.; Dübel, S.; Müller-Tiemann, B.; Trautwein, M. Evaluation of human pancreatic RNase as effector molecule in a therapeutic antibody platform. *mAbs* **2014**, *6*, 367–380. [CrossRef]

32. Sun, M.; Sun, L.; Sun, D.; Zhang, C.C.; Li, M. Targeted delivery of immuno-RNase may improve cancer therapy. *Cancer Cell Int.* **2018**, *18*, 58. [CrossRef] [PubMed]

33. Wawrzynczak, E.J.; Henry, R.V.; Cumber, A.J.; Parnell, G.D.; Derbyshire, E.J.; Ulbrich, N. Biochemical, cytotoxic and pharmacokinetic properties of an immunotoxin composed of a mouse monoclonal antibody Fib75 and the ribosome-inactivating protein α-sarcin from *Aspergillus giganteus*. *Eur. J. Biochem.* **1991**, *196*, 203–209. [CrossRef] [PubMed]

34. Rathore, D.; Batra, J.K. Construction, expression and characterization of chimaeric toxins containing the ribonucleolytic toxin restrictocin: Intracellular mechanism of action. *Biochem. J.* **1997**, *324*, 815–822. [CrossRef] [PubMed]

35. Rathore, D.; Nayak, S.K.; Batra, J.K. Overproduction of fungal ribotoxin α-sarcin in Escherichia coli: Generation of an active immunotoxin. *Gene* **1997**, *190*, 31–35. [CrossRef]

36. Carreras-Sangrà, N.; Tomé-Amat, J.; García-Ortega, L.; Batt, C.A.; Oñaderra, M.; Martínez-del-Pozo, A.; Gavilanes, J.G.; Lacadena, J. Production and characterization of a colon cancer-specific immunotoxin based on the fungal ribotoxin α-sarcin. *Protein Eng. Des. Sel.* **2012**, *25*, 425–435. [CrossRef] [PubMed]

37. Tomé-Amat, J.; Herrero-Galán, E.; Oñaderra, M.; Martínez-del-Pozo, A.; Gavilanes, J.G.; Lacadena, J. Preparation of an engineered safer immunotoxin against colon carcinoma based on the ribotoxin hirsutellin A. *FEBS J.* **2015**, *282*, 2131–2141. [CrossRef] [PubMed]

38. Jones, T.D.; Heam, A.R.; Holgate, R.G.; Kozub, D.; Fogg, M.H.; Carr, F.J.; Baker, M.P.; Lacadena, J.; Gehlsen, K.R. A deimmunised form of the ribotoxin, α-sarcin, lacking CD4+ T cell epitopes and its use as an immunotoxin warhead. *Protein Eng. Des. Sel.* **2016**, *29*, 531–540. [CrossRef]

39. Tomé-Amat, J.; Ruiz-de-la-Herrán, J.; Martínez-del-Pozo, Á.; Gavilanes, J.G.; Lacadena, J. α-sarcin and RNase T1 based immunoconjugates: The role of intracellular trafficking in cytotoxic efficiency. *FEBS J.* **2015**, *282*, 673–684. [CrossRef]

40. Tome-Amat, J.; Olombrada, M.; Ruiz-de-la-Herrán, J.; Pérez-Gómez, E.; Andradas, C.; Sánchez, C.; Martínez, L.; Martínez-Del-Pozo, Á.; Gavilanes, J.G.; Lacadena, J. Efficient in vivo antitumor effect of an immunotoxin based on ribotoxin α-sarcin in nude mice bearing human colorectal cancer xenografts. *Springerplus* **2015**, *4*, 168. [CrossRef]

41. Olmo, N.; Turnay, J.; González de Buitrago, G.; de Silanes, I.L.; Gavilanes, J.G.; Lizarbe, M.A. Cytotoxic mechanism of the ribotoxin α-sarcin. Induction of cell death via apoptosis. *Eur. J. Biochem.* **2001**, *268*, 2113–2123. [CrossRef]

42. Lacadena, J.; Alvarez-García, E.; Carreras-Sangrà, N.; Herrero-Galán, E.; Alegre-Cebollada, J.; García-Ortega, L.; Oñaderra, M.; Gavilanes, J.G.; Martínez-del-Pozo, A. Fungal ribotoxins: Molecular dissection of a family of natural killers. *FEMS Microbiol. Rev.* **2007**, *1*, 212–237. [CrossRef] [PubMed]

43. Olombrada, M.; Lázaro-Gorines, R.; López-Rodríguez, J.C.; Martínez-del-Pozo, A.; Oñaderra, M.; Maestro-López, M.; Lacadena, J.; Gavilanes, J.G.; García-Ortega, L. Fungal ribotoxins: A Review of Potential Biotechnological Applications. *Toxins* **2017**, *9*, 71. [CrossRef] [PubMed]

44. Weldon, J.E.; Pastan, I. A guide to taming a toxin-recombinant immunotoxins constructed from *Pseudomonas* exotoxin A for the treatment of cancer. *FEBS J.* **2011**, *278*, 4683–4700. [CrossRef] [PubMed]

45. Weng, A.; Thakur, M.; von Mallinckradt, B.; Beceren-Braun, F.; Gilabert-Oriol, R.; Wiesner, B.; Eickhart, J.; Böttger, S.; Melzig, M.F.; Fuchs, H. Saponins modulate the intracellular trafficking of protein toxins. *J. Control. Release* **2012**, *164*, 74–86. [CrossRef] [PubMed]

46. Pasetto, M.; Antignani, A.; Ormanoglu, E.; Buehler, E.; Guha, R.; Pastan, I.; Martin, S.E.; FitzGerald, D.J. Whole-genome RNAi screen highlights components of the endoplasmic reticulum/Golgi as a source of resistance to immunotoxin-mediated cytotoxicity. *Proc. Natl. Acad. Sci. USA* **2015**, *112*, E1135–E1142. [CrossRef] [PubMed]

47. Johannes, L.; Decaudin, D. Protein toxins: Intracellular trafficking for targeted therapy. *Gene Ther.* **2005**, *12*, 1360–1368. [CrossRef] [PubMed]

48. Ritter, G.; Cohen, L.S.; Nice, E.C.; Catimel, B.; Burgess, A.W.; Moritz, R.L.; Ji, H.; Heath, J.K.; White, S.J.; Welt, S.; et al. Characterization of posttranslational modifications of human A33 antigen, a novel palmitoylated surface glycoprotein of human gastrointestinal epithelium. *Biochem. Biophys. Res. Commun.* **1997**, *236*, 682–686. [CrossRef] [PubMed]

49. Pereira-Fantini, P.M.; Judd, L.M.; Kalantzis, A.; Peterson, A.; Ernst, M.; Heath, J.K.; Giraud, A.S. A33 antigen-deficient mice have defective colonic mucosal repair. *Inflamm. Bowel Dis.* **2010**, *16*, 604–612. [CrossRef] [PubMed]

50. Yoshida, H. The ribonuclease T1 family. *Methods Enzymol.* **2001**, *341*, 28–41.

51. Goyal, A.; Batra, J.K. Inclusion of a furin-sensitive spacer enhances the cytotoxicity of ribotoxin restrictocin containing recombinant single-chain immunotoxins. *Biochem. J.* **2000**, *345*, 247–254. [CrossRef]

52. Tortorella, L.L.; Pipalia, N.H.; Mukherjee, S.; Pastan, I.; Fitzgerald, D.; Maxfiled, F.R. Efficiency of Immunotoxin Cytotoxicity Is Modulated by the Intracellular Itinerary. *PLoS ONE* **2012**, *7*, e47320. [CrossRef] [PubMed]

53. Weldon, J.E.; Scarzynski, M.; Therres, J.A.; Ostovitz, J.R.; Zhou, H.; Kreitman, R.J.; Pastan, I. Designing the furin-cleavable linker in recombinant immunotoxins based on Pseudomonas exotoxin A. *Bioconjug. Chem.* **2015**, *26*, 1120–1128. [CrossRef] [PubMed]

54. Mutter, N.L.; Soskine, M.; Huang, G.; Alburquerque, I.S.; Bernardes, G.J.L.; Maglia, G. Modular Pore-Forming Immunotoxins with Caged Cytotoxicity Tailored by Directed Evolution. *ACS Chem. Biol.* **2018**, *13*, 3153–3160. [CrossRef] [PubMed]

55. Kaplan, G.; Lee, F.; Onda, M.; Kolyvas, E.; Bhardwaj, G.; Baker, D.; Pastan, I. Protection of the Furin Cleavage Site in Low-Toxicity Immunotoxins Based on Pseudomonas Exotoxin A. *Toxins* **2016**, *8*, 217. [CrossRef] [PubMed]

56. Kaplan, G.; Mazor, R.; Lee, F.; Jang, Y.; Leshem, Y.; Pastan, I. Improving the In Vivo Efficacy of an Anti-Tac (CD25) Immunotoxin by Pseudomonas Exotoxin A Domain II Engineering. *Large Mol. Ther.* **2018**. [CrossRef] [PubMed]

57. Meng, P.; Dong, Q.C.; Tan, G.G.; Wen, W.H.; Wang, H.; Zhang, G.; Wang, Y.Z.; Jing, Y.M.; Wang, C.; Qin, W.J.; et al. Anti-tumor effects of a recombinant antiprostate specific membrane antigen immunotoxin against prostate cancer cells. *BMC Urol.* **2017**, *17*, 14. [CrossRef] [PubMed]

58. Schapiro, F.B.; Soe, T.T.; Mallet, W.G.; Maxfield, F.R. Role of Cytoplasmic Domain Serines in Intracellular Trafficking of Furin. *Mol. Biol. Cell* **2014**, *15*, 2884–2894. [CrossRef]

59. Wise, R.J.; Barr, P.J.; Wong, P.A.; Kiefer, M.C.; Brake, A.J.; Kaufman, R.J. Expression of a human proprotein processing enzyme: Correct cleavage of the von Willebrand factor precursor at a paired basic amino acid site. *Proc. Natl. Acad. Sci. USA* **1990**, *87*, 9378–9382. [CrossRef]

60. Van de Ven, W.J.; Creemers, J.W.; Roebroek, A.J. Furin: The prototype mammalian subtilisin-like proprotein-processing enzyme. Endoproteolytic cleavage at paired basic residues of proproteins of the eukaryotic secretory pathway. *Enzyme* **1991**, *45*, 257–270. [CrossRef]

61. Mattia, A.; Merker, R. Regulation of probiotic substances as ingredients in foods: Premarket approval or "generally recognized as safe" notification. *Clin. Infect. Dis. Off. Publ. Infect. Dis. Soc. Am.* **2008**, *46*, S115–S118. [CrossRef]

62. Bader, O.; Krauke, Y.; Hube, B. Processing of predicted substrates of fungal Kex2 proteinases from *Candida albicans*, *C. glabrata*, *Saccharomyces cerevisiae* and *Pichia pastoris*. *BMC Microbiol.* **2008**, *8*, 116. [CrossRef] [PubMed]

63. Martínez-del-Pozo, A.; Gasset, M.; Oñaderra, M.; Gavilanes, J.G. Conformational study of the antitumor protein α-sarcin. *Biochim. Biophys. Acta* **1988**, *953*, 280–288. [CrossRef]

64. Pace, C.N.; Heinemann, U.; Hahn, U.; Saenger, W. Ribonuclease T1: Structure, function and stability. *Angew. Chem. Int. Ed. Engl.* **1991**, *30*, 343–454. [CrossRef]

65. Pérez-Cañadillas, J.M.; Santoro, J.; Campos-Olivas, R.; Lacadena, J.; del Pozo, A.M.; Gavilanes, J.G.; Rico, M.; Bruix, M. The highly refined solution structure of the cytotoxic ribonuclease alpha-sarcin reveals the structural requirements for substrate recognition and ribonucleolytic activity. *J. Mol. Biol.* **2000**, *299*, 1061–1073. [CrossRef] [PubMed]

66. Carmichael, J.A.; Power, B.E.; Garrett, T.P.J.; Yazaki, P.J.; Shively, J.E.; Raubischek, A.A.; Wu, A.M.; Hudson, P.J. The crystal structure of an anti-CEA scFv diabody assembled from T84.66 scFvs in V(L)-to-V(H) orientation: Implications for diabody flexibility. *J. Mol. Biol.* **2003**, *326*, 341–351. [CrossRef]

67. Wilkinson, I.C.; Hall, C.J.; Veverka, V.; Shi, J.Y.; Muskett, F.W.; Stephens, P.E.; Taylor, R.J.; Henry, A.J.; Carr, M.D. High resolution NMR-based model for the structure of a scFv-IL-1beta complex: Potential for NMR as a key tool in therapeutic antibody design and development. *J. Biol. Chem.* **2009**, *284*, 31928–31935. [CrossRef] [PubMed]

68. Oñaderra, M.; Mancheño, J.M.; Gasset, M.; Lacadena, J.; Schiavo, G.; del Pozo, A.M.; Gavilanes, J.G. Translocation of α-sarcin across the lipid bilayer of asolectin vesicle. *Biochem. J.* **1993**, *295*, 221–225. [CrossRef]

69. Gruenberg, J. Lipids in endocytic membrane transport and sorting. *Curr. Opin. Cell Biol.* **2003**, *15*, 382–388. [CrossRef]

70. Bissing, C.; Gruenberg, J. Lipid sorting and multivesicular endosome biogenesis. *Cold Spring Harb. Perspect. Biol.* **2013**, *5*, a016816.

71. Van Meer, G.; Voelker, D.R.; Feigenson, G.W. Membrane lipids: Where they are and how they behave. *Nat. Rev. Mol. Cell Biol.* **2008**, *9*, 112–124. [CrossRef]

72. Zhao, P.; Liu, F.; Zhang, B.; Liu, X.; Wang, B.; Gong, J.; Yu, G.; Ma, M.; Lu, Y.; Sun, J.; et al. MAIGO2 is involved in abscisic acid-mediated response to abiotic stresses and Golgi-to-ER retrograde transport. *Physiol. Plant.* **2013**, *148*, 246–260. [CrossRef] [PubMed]

73. Martínez-Ruiz, A.; García-Ortega, L.; Kao, R.; Lacadena, J.; Oñaderra, M.; Mancheño, J.M.; Davies, J.; del Pozo, A.M.; Gavilanes, J.G. RNAse U2 and α-sarcin: A study of relationships. *Methods Enzimol.* **2001**, *341*, 335–351.

74. Blanco-Toribio, A.; Lacadena, J.; Nuñez-Prado, N.; Álvarez-Ciénfuegos, A.; Villate, M.; Compte, M.; Sanz, L.; Blanco, F.J.; Álvarez-Vallina, L. Efficient production of single-chain fragment variable-based N-terminal trimerbodies in *Pichia pastoris*. *Microb. Cell Factories* **2014**, *13*, 116–124. [CrossRef] [PubMed]

75. Lázaro-Gorines, R.; Ruiz-de-la-Herrán, J.; Navarro, R.; Sanz, L.; Álvarez-Vallina, L.; Martínez-del-Pozo, A.; Gavilanes, J.G.; Lacadena, J. A novel Carcinoembryonic Antigen (CEA)-Targeted Trimeric Immunotoxin shows significantly enhanced Antitumor Activity in Human Colorectal Cancer Xenografts. *Sci. Rep.* **2019**, *9*, 11680. [CrossRef] [PubMed]

76. Lacadena, J.; Martínez del Pozo, A.; Martínez-Ruiz, A.; Pérez-Cañadillas, J.M.; Bruix, M.; Mancheño, J.M.; Oñaderra, M.; Gavilanes, J.G. Role of histidine-50, glutamic acid-96, and histidine-137 in the ribonucleolytic mechanism of the ribotoxin alpha-sarcin. *Proteins* **1999**, *37*, 474–484. [CrossRef]

77. Kao, R.; Martínez-Ruiz, A.; Martínez del Pozo, A.; Crameri, R.; Davies, J. Mitogillin and related fungal ribotoxins. *Methods Enzymol.* **2001**, *341*, 324–335. [PubMed]

78. García-Ortega, L.; Masip, M.; Mancheño, J.M.; Oñaderra, M.; Lizarbe, M.A.; García-Mayoral, M.F.; Bruix, M.; Martínez-del-Pozo, A.; Gavilanes, J.G. Deletion of the NH2-terminal beta-hairpin of the ribotoxin alpha-sarcin produces a nontoxic but active ribonuclease. *J. Biol. Chem.* **2002**, *277*, 18632–18639. [CrossRef]

79. Steyaert, J. A decade of protein engineering on ribonuclease T1-atomic dissection of the enzyme-substrate interactions. *Eur. J. Biochem.* **1997**, *247*, 1–11. [CrossRef]

80. Yoshimori, T.; Yamamoto, A.; Moriyama, Y.; Futai, M.; Tashiro, Y. Bafilomycin A1, a Specific Inhibitor of Vacuolar-type H$^+$-ATPase, Inhibits Acidification and Protein Degradation in Lysosomes of Cultured Cell. *J. Biol. Chem.* **1991**, *266*, 17707–17712.

81. Yeung, T.M.; Shaan, C.G.; Wilding, J.L.; Muschel, R.; Bodmer, W.F. Cancer stem cells from colorectal cancer-derived cell lines. *Proc. Natl. Acad. Sci. USA* **2009**, *107*, 3722–3727. [CrossRef]

82. Nazarewicz, R.R.; Salazar, G.; Patrushev, N.; San Martin, A.; Hilenski, L.; Xiong, S.; Alexander, R.W. Early endosomal antigen 1 (EEA1) is an obligate scaffold for angiotensin II-induced, PKC-α-dependent Akt activation in endosomes. *J. Biol. Chem.* **2011**, *286*, 2886–2895. [CrossRef] [PubMed]

83. Anderson, R.G.; Orci, L. A view of acidic intracellular compartments. *J. Cell Biol.* **1988**, *106*, 539–543. [CrossRef] [PubMed]

84. Zhao, W.; Chen, T.L.; Vertel, B.M.; Colley, K.J. The CMP-syalic acid transporter is localized in the medial-trans Golgi and possesses two specific endoplasmic reticulum export motifs in its carboxyl-terminal cytoplasmic tail. *J. Biol. Chem.* **2006**, *281*, 31106–31118. [CrossRef] [PubMed]

85. Zinchuk, V.; Zinchuk, O.; Okada, T. Quantitative Colocalization Analysis of Multicolor Confocal Immunofluorescence Microscopy Images: Pushing Pixels to Explore Biological Phenomena. *Acta Histochem. Cytochem.* **2007**, *40*, 101–111. [CrossRef] [PubMed]

86. Fujiwara, T.; Oda, K.; Yokota, S.; Takatsuki, A.; Ikehara, Y. Brefeldin A Causes Disassembly of the Golgi Complex and Accumulation of Secretory Proteins in the Endoplasmic Reticulum. *J. Biol. Chem.* **1988**, *263*, 18545–18552. [PubMed]

MDPI

St. Alban-Anlage 66

4052 Basel

Switzerland

Tel. +41 61 683 77 34

Fax +41 61 302 89 18

www.mdpi.com

Toxins Editorial Office

E-mail: toxins@mdpi.com

www.mdpi.com/journal/toxins